Project Man[illegible]it

Project Management Toolkit

The Basics for Project Success

Second Edition

Trish Melton

www.icheme.org

IChemE
heart of the process

AMSTERDAM • BOSTON • HEIDELBERG • LONDON •
NEW YORK • OXFORD PARIS • SAN DIEGO •
SAN FRANCISCO • SINGAPORE • SYDNEY • TOKYO
Butterworth-Heinemann is an imprint of Elsevier

Butterworth-Heinemann is an imprint of Elsevier
Linacre House, Jordan Hill, Oxford OX2 8DP, UK
30 Corporate Drive, Suite 400, Burlington, MA 01803, USA

First published by IChemE 2005
Second edition 2007

British Library Cataloguing in Publication Data
A catalogue record for this book is available from the British Library

Library of Congress Cataloging-in-Publication Data
A catalog record for this book is available from the Library of Congress

ISBN: 978-0-7506-8440-8

For information on all Butterworth-Heinemann publications
visit our web site at books.elsevier.com

Typeset by Charon Tec Ltd (A Macmillan Company), Chennai, India
www.charontec.com
Printed and bound in Great Britain

07 08 09 10 10 9 8 7 6 5 4 3 2 1

About the author

Trish Melton is a project and business change professional who has worked on engineering and non-engineering projects worldwide throughout her career. She works predominantly in the chemicals, pharmaceuticals and healthcare industries.

She is a chartered Chemical Engineer and a Fellow of the Institution of Chemical Engineers (IChemE), where she was the founder Chair of the IChemE Project Management Subject Group. She is a part of the Membership Committee which reviews all applications for corporate membership of the institution and in 2005 she was elected to the Council (Board of Trustees)

She is an active member of the International Society of Pharmaceutical Engineering (ISPE) where she serves on the working group in charge of updating ISPE's *Bulk Pharmaceutical Chemicals Baseline® Guide*. She is the founder and Chair of the Project Management Community of Practice formed in 2005. She has presented on various subjects at ISPE conferences including project management, quality risk management and lean manufacturing and has also supported ISPE as the conference leader for project management and pharmaceutical engineering conferences. She is also the developer and lead trainer for ISPE's Project Management Training Course. In 2006 the UK Affiliate recognized Trish's achievements when she was awarded their Special Member Recognition Award.

Trish is the Managing Director of MIME Solutions Ltd., an engineering and management consultancy providing project management, business change management, regulatory, and GMP consulting for pharmaceutical, chemical, and healthcare clients.

Within her business, Trish is focused on the effective solution of business challenges and these inevitably revolve around some form of project: whether a capital project, an organizational change programme or an interim business solution. Trish uses project management on a daily basis to support the identification of issues for clients and implementation of appropriate, sustainable solutions.

Good project management equals good business management and Trish continues to research and adapt best practice project management in a bid to develop, innovate and offer a more agile approach.

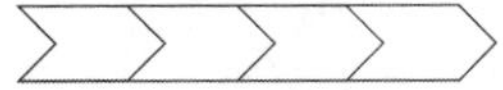

About the Project Management Essentials series

The Project Management Essentials series comprises four titles written by experts in their field and developed as practical guidelines, suitable as both university textbooks and refreshers/additional learning for practicing Project Managers.

- Project Management Toolkit: The Basics for Project Success
- Project Benefits Management: Linking Projects to the Business
- Real Project Planning: Developing a Project Delivery Strategy
- Managing Project Delivery: Maintaining Control & Achieving Success

This first book provides a general overview with the subsequent titles supplementing the skills and knowledge gained and expanding the toolkit. The books in the series are supported by an accompanying website (www.icheme.org/projectmanagement), which delivers blank tool templates for the reader to download for personal use.

Foreword

This book has become a reality for a number of reasons:

- As an experienced Project Manager I realized that increasingly I was dealing with customers, sponsors and Project Team members who had no project management experience. *Project Management Toolkit* is a direct response to this. For many years I have used the material contained in this book to develop and train 'line' managers, in particular, those who needed the basic skills to successfully deliver a project they had just been 'handed'.
- As the found Chair of the IChemE Project Management Subject Group (PMSG) and then more recently a part of the Continuous Professional Development (CPD) and Publications Sub-groups it was evident that there wasn't a good entry-level book for those wanting to develop project management skills at the start of their project management careers.

Project Management Toolkit is an entry-level book and the first in a new series of project management books developed by members of the IChemE PMSG. The three subsequent books in the series are at a more detailed level to supplement the skills and knowledge developed in *Project Management Toolkit* (see page v for more information).

This second edition of *Project Management Toolkit* has responded to the excellent feedback from readers and reviewers and provides some more 'real' examples of how the toolkit has been used. Additionally these new case studies take the reader through every stage of the project and through every applicable tool.

Although this book is primarily written from the perspective of engineering projects within the process industries, experiences from both outside of this industry and within different types of projects have been used. Based on the feedback from some of the Project Managers who have used this book the two new case studies cover two very different types of project: the capital engineering project and the business change project (usually revenue expenditure) — demonstrating the generic use of the toolkit over a wide range of industries and project types.

The *Toolkit* is specific enough to support Engineering Managers in the delivery of projects within the process industries, yet generic enough to support

- the Research and Development Managers in developing or launching a new product;
- Business Managers in transforming a business area
- IT Managers in delivering a new computer system

Project management is about people, and this *Toolkit* emphasizes the criticality of the management of the 'soft' side of projects — the people whose lives may change as a result of a project, the Project Team members who are key to effective delivery and the sponsors and organizational stakeholders who ensure, with the Project Manager, that 'no project is an island'.

Acknowledgements

When pulling together ideas and experiences into a 'book' you become really aware of all the friends, family and colleagues who have helped you along the way to a greater or smaller extent. You are equally aware that to individually acknowledge everyone becomes impossible; therefore, I have picked out a few key ones. . . .

To Andrew, my husband, without whom I would never eat! Life is a project and Andrew is, and will always be, my critical path.

To all my 'test' audiences over the years who have enthusiastically used the tools to challenge whether they are doing the 'right' project and then whether they are doing it 'right' — the fact that they have used the tools on a broad spectrum of project types and sizes within many different industries has reinforced my view that at some level project management is based on generic principles.

Every good Project Manager needs a team — I am fortunate in having a wide team of colleagues — particularly Gillian Lawson and Peter Iles-Smith, my Project Management Subject Group (PMSG) colleagues who, with me, are the IchemE Editorial Team for the Project Management Series of Books. I also want to thank Arnold Black and Mike Adams who have provided an invaluable 'sounding board' for my PM thoughts over the years we have worked together on the PMSG Committee.

Finally you always need 'live' Project Teams to test new project management ideas, tools and processes — AstraZeneca Transformation Projects Group (led by Paul Burke) provided this role during the development of the first edition. Without them I may have finished the book sooner; however, I just wouldn't have had as much fun! I'm thankful for their challenge and their enthusiasm.

I also want to thank Paul Burke, Jeff Wardle and Bill Wilson for continuing to use and test the tools in this book and allowing me to use example data for the second edition based on their business change projects within the Strategy Planning and Change Management group within the AZ UK Business Services organization.

Author's Note: Although all the case studies presented in this book are based on real experiences they have been suitably altered so as to maintain complete confidentiality.

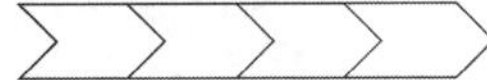

How to use this book

When you pick up this book I am hoping that before you delve into the content you will start by glancing here.

The structure for this book is based on the concept that every project goes through four value-added stages — these are described in Chapter 1. Each stage then becomes the subject of its own chapter (Chapters 3–6).

Chapters 1 and 2 are general introductions and overviews to project management which can be read at any time to refresh basic concepts that apply to every stage of a project.

Chapters 3–6 are the 'core chapters' made up of the following generic sections:

- Introduction of basic concepts particular to a project stage.
- Presentation of new tools and how to use them.
- Demonstration of tool use through short case studies.
- Handy hints and further reading.

Each core chapter can 'stand-alone' allowing you to dip into any stage and within that stage, any tool or case study. For each tool the same structure is followed:

- The tool is introduced within the context of the stage.
- The tool is explained through use of a completed template plus additional notes.
- The use of the tool within a project context is detailed.

The blank tool templates are available, in pdf format, on a password protected area within IChemE's web site: www.icheme.org/projectmanagement. The password allowing access to the tool templates is PMTOOLS. The actual format of the pdf cannot be changed but project data, as required by the tool, can be either input electronically or by hand on printed copies.

Some of the tools are more advanced than others and may be more applicable to larger, more complex projects. You need to decide what tools you want to use, how you want to use them and then adapt them for use on your projects.

At the end of each core chapter there is a brief section on further reading, which highlights the specific book in the PM Essentials series where the project stage is covered to a greater depth. Additionally the following web sites can be used as a source of further information on project management (see Footnote 1):

- IChemE Project Management Subject Group (PMSG) — www.icheme.org/pmsg — the vision of this group is to facilitate networking between Project Managers and aspiring Project Managers, of any discipline, within the process industries; to promote 'best practice' project management within the process industries and to be the 'voice of project management' for the IChemE.

Footnote 1: All information regarding project management associations and institutes is available to the public via their web sites and the author/publisher of this book does not take any liability for its veracity.

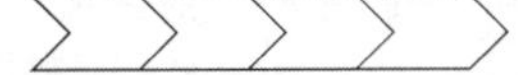

- The Association of Project Management (APM) — www.apm.org.uk — is an independent professional body based in Europe. APM's key objective are to develop and promote project management across all sectors of industry and beyond. A key resource within the APM is the APM Body of Knowledge.
- Project Management Institute (PMI) — www.pmi.org — is a global project management institute focused on the needs of Project Managers worldwide. It has over 200,000 members representing 125 countries and offers professional development support via its certification programme, education events and Project Management Body of Knowledge.
- International Project Management Association (IPMA) — www.ipma.ch — is the world's oldest project management association. It is an international network of national project management societies and is able to represent these national societies at international level. IPMA actively promotes 'the importance of efficient, enterprise-wide project management competencies to organizations' and has a certification programme supported by educational events, links to academia and research.
- International Society of Pharmaceutical Engineering Project Management Community of Practice (ISPE PMCoP) — www.ispe.org/pmcop — the aim of this group is to be a dynamic forum for professionals working within the pharma industry who have an active interest in promoting continuous improvement in project management and also to create a body of knowledge specific to the professional needs of its members. The group encourages discussion of areas of common interest and provides information of relevance to members through educational events.

Chapter 7 and 8 are new to this second edition and are provided as complete case studies taking the reader through every stage of two selected projects:

- The Pharma Facility Project — an engineering project example highlighting some of the challenges in delivering a capital project within a highly regulated environment.
- The Business Change Project — an organizational design project highlighting some of the issues associated with the delivery and sustainability of changes which impact the way in which the business operates.

And remember . . .

There are many ways to complicate project management – some valid and others not. This book introduces some basic tools so that at each 'value-added' stage the aspiring Project Manager can focus on delivering value.

- The greatest compliment made to me by a client was **'did we need you?'**
- The greatest compliment made to me by a project sponsor was **'why were there no problems?'**

It used to be that the best Project Managers were those that solved the big crises, like heroes putting out the fire whilst saving the business. . . . Today this is not success — a great Project Manager will not allow any inferno — he'll smell the smoke way before ignition and put it out with the help and support of his team and his stakeholders.

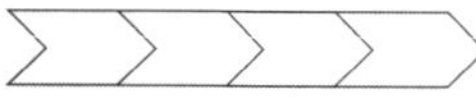

Contents

1 Introduction

Project Management Toolkit is a practical handbook for both career Project Managers and managers involved with projects intermittently throughout their career.

Organizations are realizing the impact that projects, and therefore project management, can have on their success.

A project used to be one mechanism that organizations used to deliver benefits, now organizations are managed by project; this has meant the development of project management competency within the organization, the Project Team(s) and the individual Project Manager(s).

This book provides a simple tool based resource for each key 'value-added' stage in a project and is intended for use by anybody involved in projects and could form a basis for an organizational project management system (Figure 1-1).

At the end of each chapter 'handy hints' are given as well as details of the specific book in the IChemE project management series where more information can be sought.

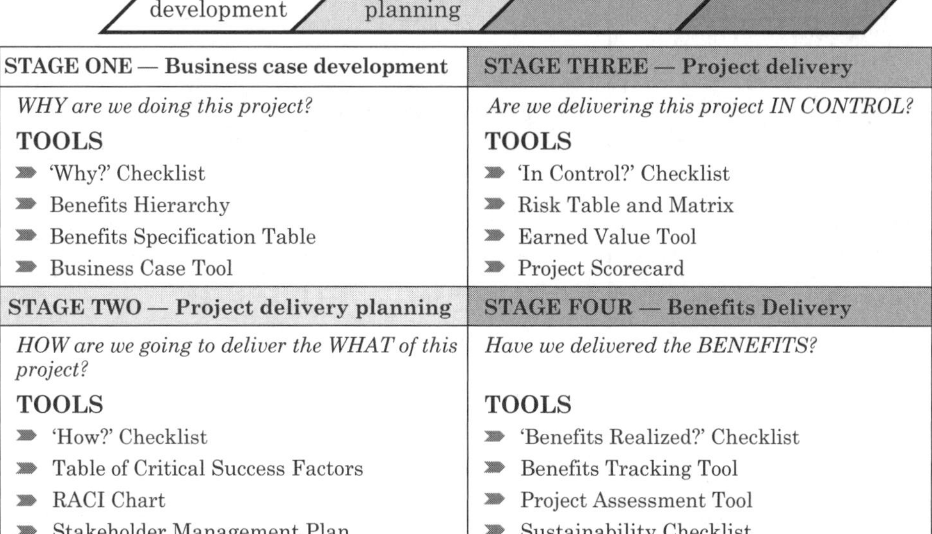

STAGE ONE — Business case development	STAGE THREE — Project delivery
WHY are we doing this project?	*Are we delivering this project IN CONTROL?*
TOOLS	**TOOLS**
➤ 'Why?' Checklist	➤ 'In Control?' Checklist
➤ Benefits Hierarchy	➤ Risk Table and Matrix
➤ Benefits Specification Table	➤ Earned Value Tool
➤ Business Case Tool	➤ Project Scorecard
STAGE TWO — Project delivery planning	**STAGE FOUR — Benefits Delivery**
HOW are we going to deliver the WHAT of this project?	*Have we delivered the BENEFITS?*
TOOLS	**TOOLS**
➤ 'How?' Checklist	➤ 'Benefits Realized?' Checklist
➤ Table of Critical Success Factors	➤ Benefits Tracking Tool
➤ RACI Chart	➤ Project Assessment Tool
➤ Stakeholder Management Plan	➤ Sustainability Checklist
➤ Control Specification Table	

Figure 1-1 *Project Management Toolkit* overview

Case studies in each chapter illustrate the use of each tool. Blank tool templates are available via the internet at www.icheme.org/projectmanagement for the readers of this book to use.

Aims

The primary aim of this book is to develop a practical, usable resource which can be picked up and used on 'day 1'. It provides the reader with education, tools and the confidence to successfully manage projects.

Figure 1-2 shows an input–process–output diagram:

- Inputs — lists the inputs to the development of this book.
- Process — summarizes the contents of this book.
- Outputs — lists the outputs from this book from the perspective of the reader.

Project Management Toolkit introduces:

- Simple but effective tools which can be used to support the Project Manager in increasing the 'certainty of outcome'.
- A pragmatic process to the development of your project, from the early stages of idea development through to the delivery of the benefits.

Apart from the tools and the processes, this book also aims to define more clearly the role of the Project Manager in today's projects and the increasing importance of generic project management competency within an organization.

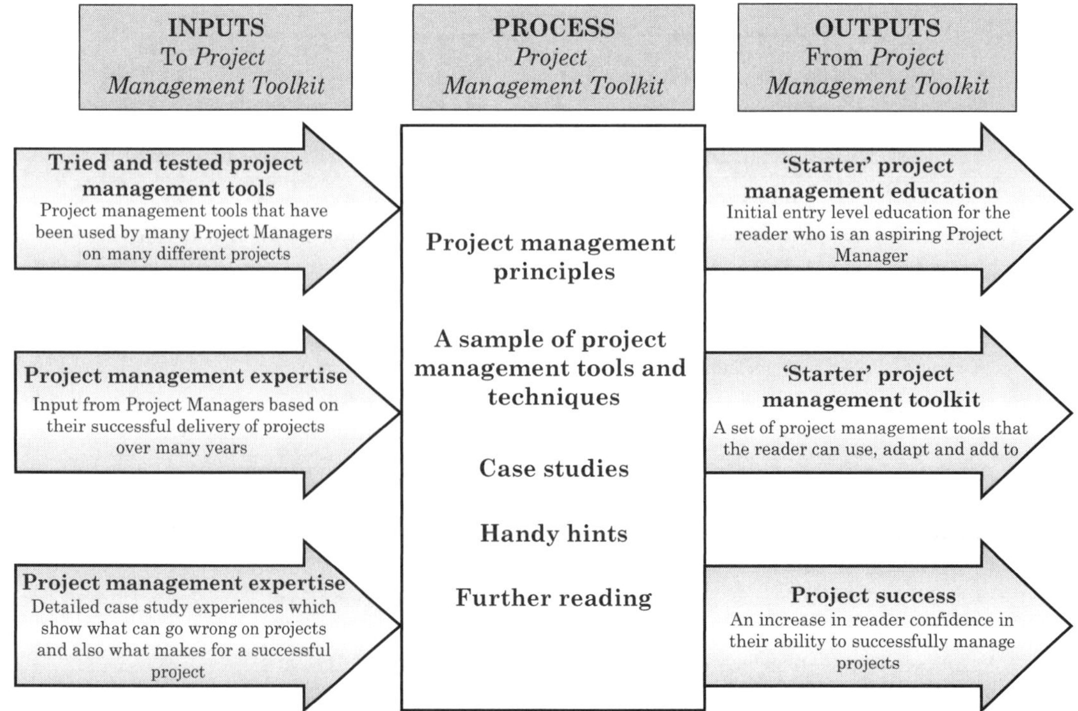

Figure 1-2 This book's IPO (input–process–output)

Although this book is primarily aimed at Project Managers within the process industries it is equally applicable to Project Managers in other disciplines because project management is a generic organizational competency which can be used:

- Within *any type* of project — by use of transferable project management skills and knowledge.
- Within *any type* of organization — by adding strategic business value to an organization.

Management by project

Each year millions of pounds are spent around the globe delivering projects. Therefore effective project management is critical for today's organizations.

Consider the organizational impact of:

- Delivering a project late.
- Delivering a project over budget.
- Delivering a project which doesn't meet scope requirements.

For some projects the impact of not delivering within these three basic parameters can have disastrous effects on an organization.

Delivering a project late

Some projects have a defined and fixed target completion date; if this date is missed then the organization may not be able to realize the benefits. For example a manufacturing facility which will support the launch of a new pharmaceutical product for the treatment of respiratory disease needs to be complete in time for the winter launch of the drug and certainly before the launch of a competitor drug.

Delivering a project over budget

A project budget is a key part of the 'organizational contract'; the benefits which will be realized are directly related to the investment monies approved. For example a project to automate a production process is approved so that the production capacity increases; if the investment is greater than budgeted then the organization will not realize the expected benefits.

Delivering a project that doesn't meet the scope requirements (quantity, quality, functionality)

A project delivers a specific amount of scope at a specified level of quality with certain functional requirements; if this is not delivered then the completed project may not be able to deliver the anticipated benefits. For example a project to improve production efficiency if not capable of enabling the business benefits to be realized, and for those benefits to be tracked, cannot be considered as successfully supporting the organization.

Organizational project management

Recent research by the Engineering Construction Industry Association (ECIA) was able to demonstrate the financial impact of good and poor project management:

- Effective use of best practice project management yielded an average cost saving of 5–10%.
- Poor use of project management gave an average cost growth of 10%.

Apart from showing that the use of best practice project management was able to decrease the average percentage cost growth, the data also showed a decrease in variability — the use of best practice project management increased the ability to forecast the outcome. It could be said that:

- Excellent Project Managers have the capability to bring projects in on time and within budget — average or poor Project Managers may not!

Recent management theorists, and project management practitioners, have proposed a new cultural paradigm that relies on project management competency as a core skill for an organization:

- Organizations are moving from *managing projects* to *management by projects.*

A change in the way these organizations do business relies on project management competence at all levels in the organization as a key success factor, that is to say within Project Teams and the business management team at all levels.

Organizations are realizing the impact that projects, and therefore project management, can have on their success:

- A project used to be one mechanism that organizations used to deliver benefits — now projects & project management are integral to normal business operation.

This has meant the development of two further facets of project management competency:

- Organizational project management excellence.
- Project Team excellence.

Therefore the competency of the Project Manager is not the only determining factor. It is suggested that if you have a project focused organization, with excellent technical ability and effective teamwork then you still need an excellent Project Manager to achieve an excellent outcome — to exceed your objectives.

Figure 1-3 outlines the various elements that a project-focused organization should consider:

- Organizational support system.
- Project Manager support system.
- Project management processes.

Project Office
Corporate Support
Organization Competency Framework
Knowledge Framework
ORGANIZATIONAL SUPPORT SYSTEM
PROJECT MANAGER SUPPORT SYSTEM
PROJECT MANAGEMENT PROCESSES
Project Management Community
Tools and Techniques
Competency framework
Career ladder
Project management principles and standards
Team competency framework

Figure 1-3 An organizational approach to project management

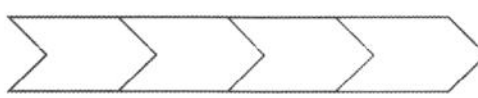

Organizational support system

This is a system to support the management of the organization 'by project'. It supports the development of a generic project management competency for all individuals in that organization.

Project Manager support system

This is a system to support the development of project management excellence within that organization — within the career Project Managers. Career ladders may be linked to formally recognized professional qualifications, to specific education providers or to specific project management institutions such as the APM (Association of Project Management).

Project management processes

This is a collection of the appropriate project management tools for that organization and is thus support for the entire organization, including career Project Managers. The overall roadmap for any project would be defined here.

This book introduces basic tools and techniques, which could form the basis for the development of robust organizational project management processes.

The Project Manager

Project Managers are those individuals who remain accountable for the achievement of the project objectives and who also ensure alignment of the project objectives with the business objectives via use of an organizational project sponsor. This role is described in more detail in Chapter 3.

Traditionally Project Managers have focused on 'the project triangle' and all the tools, techniques and processes have been about the delivery of this triangle — the management and control of scope (quantity, quality and functionality), cost and time (Figure 1-4).

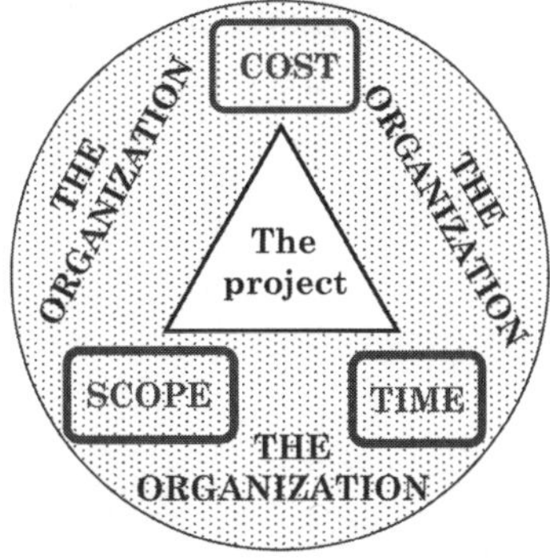

Figure 1-4 The project triangle

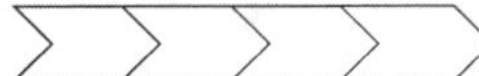

Table 1-1 Project Managers

	Career Project Manager	Temporary project involvement
The role	➤ Accountable for projects and possibly for programmes of projects ➤ Developed over a number of years with practical project experience	➤ Line Manager typically delivering one project and then returning to line management role (project is usually linked to this role) ➤ Addressing an immediate project need ➤ Potentially has limited project experience but valuable background and understanding of the organizational and/or technical issues
Development needs	➤ Structured, progressive continuing training and development ➤ In-depth expert level training ➤ Potential to have a formal project management qualification	➤ Fast introduction to project management processes, tools and techniques ➤ Guidance on 'where to get help' ➤ Coaching
Main outcome	➤ Individual experts available to an organization for professional project delivery	➤ Individual basic project management understanding ➤ Improved organizational project management competence

Now Project Managers need to understand the environment in which the project is to be delivered. As well as having a robust control strategy they must equally have a link to the organization. This has lead to an increasing awareness of 'benefits management', which is described in detail in Chapter 3. Therefore Project Managers are considered to be:

- Accountable for the achievement of the project objectives (cost, scope and time), for example through an appropriate project management control methodology.
- Responsible for the delivery of the benefits enablers (the things that allow the benefits to be realized), for example by understanding the link between the project and the business.
- Responsible for developing the Project Team and managing the project stakeholders (anyone involved with or impacted by the project) — through good people management.

Project management competence can be developed in two ways within an organization (Table 1-1).

As organizations become more aware of the importance of projects and project management they have realized the importance of project management competence.

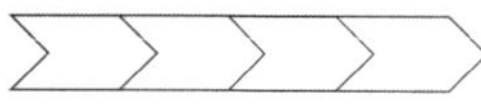

2 The project life-cycle

There are many definitions of 'a project' and there are equally many descriptions of the overall life-cycle of a project, so the aim of this brief chapter is to present one definition of each and then to look at some particular facets of the project life-cycle in readiness for the presentation of the project management toolkit in Chapters 3 to 6.

What is a project?

A project is a distinct package of scope which when delivered will enable the organization to realize a distinct package of benefits. Once the benefits are being realized sustainably the project has been integrated into the organizational processes — it has enabled some change in 'the way we do things round here'.

A project therefore has:

- A start point.
- An end point.
- A specific target to achieve.

What is the project life-cycle?

A project goes through four distinct 'value-added' stages from its start point to its end point (Figure 2-1). Each stage has its own start and end point and each has a specific target to achieve, that is to say each stage within a project is a project in its own right (Table 2-1).

Stage One: business case development

The project start point is usually an idea within the business — for example an identified need, a change to the status quo or a business requirement for survival. At this stage the project management processes should be challenging whether this is the 'right' project to be progressing.

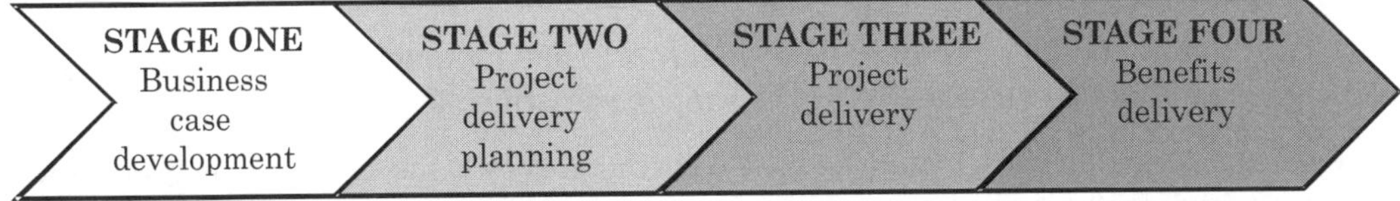

Figure 2-1 The four 'value-added' project stages

Table 2-1 Summary of the project life-cycle

	Start point	End point	Specific target
Stage One	An idea	Agreement that this is a project which meets a business need	Determine the link between the project and the business, confirm the project scope which will enable the business benefits
Stage Two	An approved project (approval to develop a plan)	An approved project delivery plan	Determine the most appropriate way to deliver the project scope to meet the business needs. Ensure that the business plans to receive the project
Stage Three	An approved project (approval to deliver)	A successfully delivered project	Deliver the appropriate project scope in control
Stage Four	A delivered project	Sustainably delivered benefits	Deliver the required business benefits sustainably. Integration of any business changes caused by the project

Stage Two: project delivery planning

This stage is all about planning and the project management processes are used to determine how to deliver the project 'right'.

Stage Three: project delivery

Effective delivery is all about the control and management of uncertainty. This stage is therefore focused on the controlled delivery — to deliver it 'right'.

Stage Four: benefits delivery

The final stage involves integrating the project into the business — allowing the project to become a part of the normal business process.

Future references to 'value-add' are reminders that 'no project is an island', that is to say a project should deliver value to its customer. The delivery of value is the link between the project and the business going on around it.

The physical link between the project and the business is usually through the relationship between the Project Manager and the project sponsor and also through the project sponsor with the customer (Table 2-2).

Why do projects fail?

There are clearly many distinct types of project (Figure 2-2) and although each may have very specific needs and targets they can be managed using the same basic principles. All projects need:

- A Project Manager and Project Team who have a shared goal — a successful project outcome.
- A business issue to solve and a business sponsor to own that issue and its successful resolution.
- A customer and end-user group who will integrate the results of the project into their daily lives.

Table 2-2 Project Manager, project sponsor and customer needs

	Project Manager	Project sponsor	Customer
Stage One	➤ To develop the idea into a scope to address the business need — a business case ➤ To agree the scope with the customer ➤ To agree the business need with the sponsor	➤ To ensure that the business needs the project ➤ To understand the business case	➤ Has an idea for a project ➤ Works with the project manager to define the quantity, quality and functionality of a scope ➤ To resolve any issues
Stage Two	➤ To develop a plan to deliver the agreed scope	➤ To ensure that the plan will enable the business needs to be met	➤ To understand how the project will deliver the required scope
Stage Three	➤ To deliver the project in control ➤ To receive support from the sponsor linked to resource requirements ➤ To receive information and decisions from the customer as required by the plan	➤ To know how the project is progressing with respect to the certainty that the business need will be met ➤ To understand any changes which impact on the ability to meet the agreed business case	➤ To know how the project is progressing with respect to the certainty that the scope will be delivered ➤ To understand any changes in scope and schedule in particular
Stage Four	➤ To support the business with any integration of change into the business	➤ To measure business benefits ➤ To ensure that all benefits are being sustainably realized	➤ To measure local benefits ➤ To maintain sustainability by integrating any business changes caused by the project, into the normal operational processes

If the 'technical' differences between projects are ignored, the reason why most projects fail can be summarized:

- No robust business case — no answer to 'why?'
- No robust statement of scope — no answer to 'what?'
- No robust project delivery plan — no answer to 'how?' or 'who?' or 'when?'
- A lack of control during delivery — no answer to 'will we succeed?'
- A lack of delivery of business benefits — no answer to 'so why did we do this project?'

The above is based on previous experiences from many sources and a summary of 'common pitfalls' is included in Table 2-3.

A major issue with most projects, though, has been the inability to successfully merge the 'hard' and 'soft' side of the projects and the associated project management.

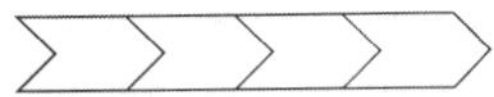

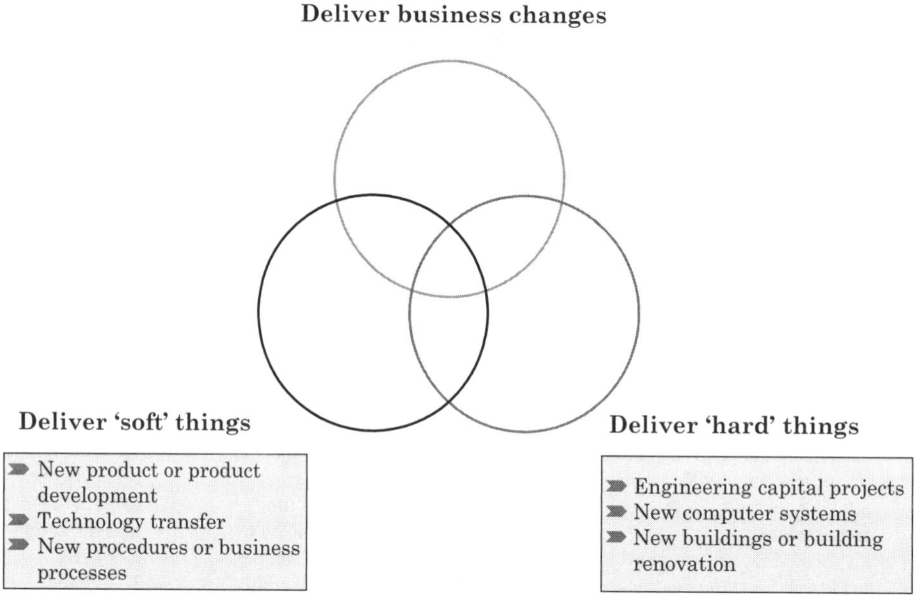

Deliver business changes

Deliver 'soft' things

- New product or product development
- Technology transfer
- New procedures or business processes

Deliver 'hard' things

- Engineering capital projects
- New computer systems
- New buildings or building renovation

Figure 2-2 Types of projects

The 'hard' and 'soft' side of project management

When 'hard' project issues are talked about the more traditional aspects of project management are usually being discussed, for example:

- The scope definition.
- Project timeline and project costs.
- Risk and issues.
- Tangible objectives (for example deliverables).
- Business benefits (for example financial savings and investment payback).
- Project controls.
- *'Getting it done.'*

When 'soft' project issues are talked about, less traditional aspects of project management are usually being discussed, for example:

- How people work together (in the project and in the business).
- Relationships and politics.
- Ownership of the project.
- Intangible objectives and benefits (for example improving team morale).
- Influencing customers.
- *'People feel OK; business feels OK.'*

Table 2-3 Common pitfalls — 'hard' and 'soft'

Common pitfall	How we can avoid this
Hard — badly defined scope	➤ All project stakeholders need to understand what the project is trying to achieve ➤ Define the critical success factors and work with these to define all sub-projects, activities and tasks required to achieve them
Soft — ignoring relationship building	➤ Projects are all about people — people in the Project Team plus all the external stakeholders ➤ Develop a project stakeholder map, understand who is involved and how it will impact them ➤ Develop a plan to manage all relationships effectively
Hard and soft — poor or no project planning	➤ Develop a project delivery plan ➤ It's never too early to ask how?, who?, what?, when?
Hard and soft — unclear business benefits	➤ Develop a robust business case in liaison with the customer — this is what the customer wants and what the scope should deliver ➤ Use the opportunity to begin to build a robust relationship
Hard — not enough project resources	➤ The lack of money, equipment or people is a critical issue and can only be resolved through effective planning and definition ➤ Work with the sponsor to get approval for the business case and delivery plan and release of the approved resources
Soft — approved resources don't materialize	➤ Work with the sponsor to resolve the issue as early as possible. He is accountable for the benefits delivery and he can't do this if the Project Manager is unable to deliver the project
Hard and soft — no 'live' risk assessment	➤ Regular risk reviews versus all of the critical success factors (CSFs) — start early and don't stop ➤ Consider 'hard' and 'soft' risks ➤ Regularly review the overall chances of project success
Hard — 'run-out' of funds	➤ Develop a robust cost plan — review versus scope (quantity, quality and functionality) and time goals ➤ Develop robust change control processes
Hard and soft — project team do not deliver	➤ Develop a team charter (vision of project success) so that the team have a clear shared goal ➤ Develop a team organization and use something like a RACI (responsible, accountable, consulted, informed) chart to ensure that roles and responsibilities are clear ➤ Use appropriate Project Team and individual measures to track performance at all levels ➤ Review team performance regularly
Hard — project is out of control	➤ Define a control plan and stick to it! ➤ Get appropriate control tools and use them pragmatically
Hard — business benefits not delivered	➤ Develop a robust business case and include both 'hard' and 'soft' benefits. Ensure that this effectively links the benefits with the project scope ➤ Define how the benefits are to be measured — the benefit metrics — and who is accountable for their delivery
Soft — lack of change sustainability	➤ Changes as a result of the project can be checked after project delivery through the use of specific sustainability checklists ➤ Ensure that user groups are involved with the project at an early enough stage so that they can get involved, give the Project Team their knowledge and support the effective delivery of the benefits from the project

The nature of project management is changing as it is realized that a project has the ability to impact the business, for example:

- A capital project to design and build a new facility is no longer just about the 'bricks and mortar'; increasingly organizations are expecting the Project Manager or project director to consider how the facility will be used, that is to say this project delivers a 'hard' thing and some business change (Figure 2-2).
- The definition of project success is no longer the delivery of the facility — it is the delivery of a facility in operation in line with business needs.

This shift in the breadth of project management impacts on how projects are delivered and the skills required to effectively deliver them.

The 'hard' side of project management is still required — projects must have a robust business case, a well defined scope and be delivered in control (with respect to time, cost and quality objectives). Typical pitfalls related to some of these 'hard' issues are described in Table 2-3 along with suggested avoidance tactics.

The 'soft' side of project management is equally required in order to effectively manage the people and the business change issues which surround every type of project:

- Any project has the potential to change the business because people, processes and organization culture are impacted by projects.
- Any project is delivered by people and that team of people needs to be managed appropriately.

Typical pitfalls related to some of these 'soft' issues are described in Table 2-3 along with suggested avoidance tactics.

For effective management of projects in today's world the 'hard' and 'soft' sides of project management need to merge (Figure 2-2) and the project management processes must do this.

This book introduces a selection of tools which will support the above. They have been chosen based on the most common 'pitfalls' seen (Table 2-3) and are therefore not exhaustive.

Remember . . .

Successful projects:

- Have a business rationale for the project — the organization needs it.
- Have well defined scope linked to cost and schedule — they know what to deliver and it's link to business benefits.
- Are delivered in control — the Project Manager has 'certainty of outcome', he knows how likely success is, and there is no fire-fighting!
- Have delivered the business benefits — through understanding the changes that are needed or caused within the business.
- Are built on sound relationships and effective people management.

3 Stage One: why?

The first value-added stage in a project involves asking 'why?'

- Why is the project being done?
- Why does the organization need this project completing?

Every project has the potential to change the business through changing how people and business processes operate (in isolation and together). This first project stage is the start. It is the start of the benefits management life-cycle (Figure 3-1).

The benefits management life-cycle should be considered as a continuous linkage between the project and the business throughout its life — that is to say throughout the four project stages. Stage One makes the initial link and is crucial in making sure that the organization progresses the right projects. It allows:

- Understanding of why the project needs to be done.
- Development of a scope which supports the development of a robust business case.
- Ensures that the project is appropriately supported and then authorized within the organization.

Stage One covers the initial life-cycle of a project — from an idea through to an approved project: determined to be the right project for the organization at that time. Within Stage One an organization may have many stage gates to progressively determine whether this is the right project (Figure 3-2). Therefore the tools contained in this chapter do need to be put into the perspective of a particular project justification and authorization process.

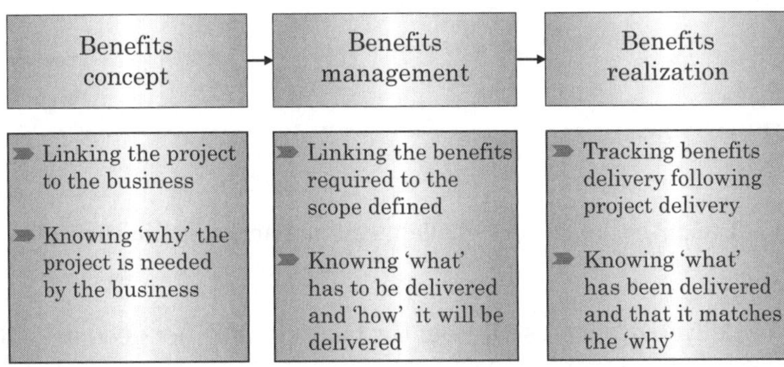

Figure 3-1 The benefits management life-cycle

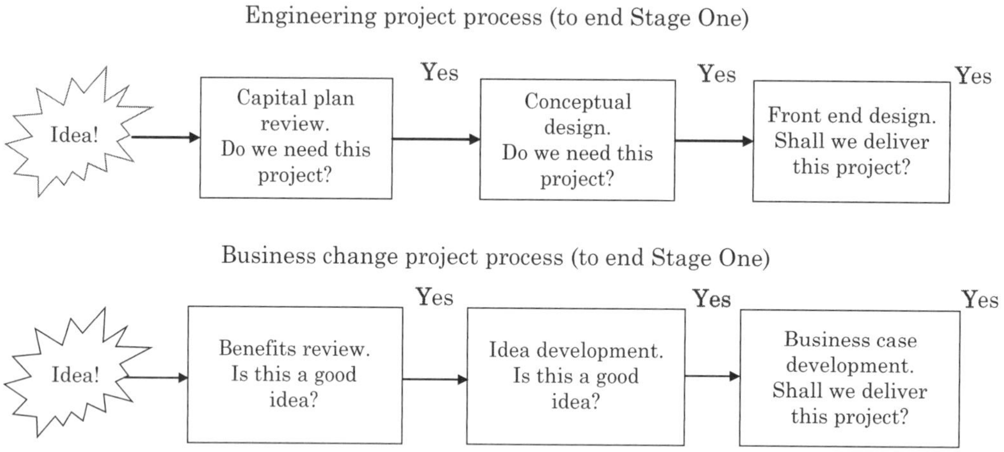

Engineering project process (to end Stage One)

Idea! → Capital plan review. Do we need this project? → Yes → Conceptual design. Do we need this project? → Yes → Front end design. Shall we deliver this project? → Yes

Business change project process (to end Stage One)

Idea! → Benefits review. Is this a good idea? → Yes → Idea development. Is this a good idea? → Yes → Business case development. Shall we deliver this project? → Yes

Figure 3-2 Example of Stage One processes

Why?

Stage One asks for a robust challenge to a project by asking 'why?' When an issue is identified this is only the symptom of the problem. By asking 'why?' the root cause of the issue can be determined and it is the root cause that the project should be addressing.

If the idea or potential project can respond positively to the challenge 'why should we do this?' you will probably already have been able to sufficiently articulate the main benefit that the project should enable for the business.

The most important question a Project Manager can ask upon receipt of a project request is 'why?' No longer can the Project Manager dodge this question with the cry 'I'm just delivering what I've been told to deliver!'

In order to deliver the project 'right' a Project Manager needs to understand that this is the 'right' project; he needs to be able to make a link between the project scope and the benefits that the scope will enable.

The following tools are just a few which can be used during the early stages of a project to provide support in asking 'why?':

- 'Why?' Checklist.
- Benefits Hierarchy.
- Benefits Specification Table.
- Business Case Tool.

Projects which do not receive this level of challenge at an early stage in their life are more likely to fail:

- The project may be delivered 'successfully' but the business may not need it — the project effectively becomes a 'white elephant'.
- The project may not deliver at all!

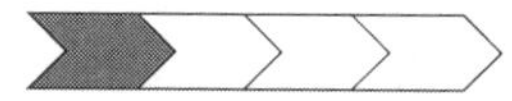

Tool: 'Why?' Checklist

If any project (at whatever stage) cannot be challenged by the questions contained in this checklist *and* have robust answers, suitably backed up, then it isn't robust enough to be delivered. This may seem a bold statement; however, the questions are basic and address the rationale for the project's existence.

This tool can also be used proactively throughout Stage One as a reminder of what should be done next — to support the robust development of an idea so that the business can approve the associated project.

The 'Why?' Checklist explained

The checklist contains five major checks:

- Sponsorship.
- Business benefits.
- Business change.
- Scope definition.
- Stage One decision.

The 'Why?' Checklist is shown in Table 3-1 with high-level guidance on how to complete each check. The following are additional, more detailed, notes to support checklist completion.

Sponsorship

The aim of the two questions used here is to determine if the project has sufficient level of support in the organization:

- Who is the sponsor?
- Has the sponsor developed an external communication plan?

The sponsor should be identified by name and by position in the organization so that it is clear that they have the appropriate level of authority (Table 3-2). This is a key part of project stakeholder management, which begins in Stage One and is then developed in more detail during Stage Two using tools such as the Stakeholder Management Plan (see page 64). The role of the sponsor, and other key stakeholders, is explained in more detail within the Stakeholder Management Plan.

At this stage it is critical that the Project Manager and project sponsor develop a joint 'contract' so that each understands the other's role — responsibilities and accountabilities. This is still a very early stage in the life of the project and it is likely that the sponsor and Project Manager will need to review this 'contract' at the start of each subsequent stage of the project.

The sponsor should also be considering how the business needs to be kept updated during the development of the idea/project. Consequently a summary communications plan detailing who should be updated and with what is a useful tool.

Business benefits

These four checks are at the heart of the 'why?' rationale and start to address benefits management. When an idea is initially thought up it is unlikely that a full business case will have been developed.

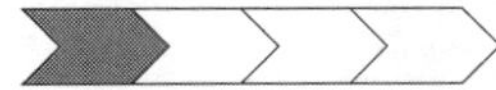

Table 3-1 The 'Why?' Checklist explained

Project Management Toolkit — 'Why?' Checklist

Project: *<insert project title>*	**Project Manager:** *<insert name>*
Date: *<insert date>*	**Page:** 1 of 1

Sponsorship

Who is the sponsor? (The person who is accountable for the delivery of the business benefits)
<insert the name of the person who is taking this role>
Has the sponsor developed an external communication plan? (How the sponsor will communicate with all stakeholders in the business)
<insert any comments on how the sponsor has/is communicating with the business>

Business benefits

Has a business case been developed?
<insert comments on the current status of the formally developed business case which supports the project>
Have all benefits been identified? (Why is the project being done?)
<insert comments on the progress of the articulation of the benefits of completing the project>
Who is the customer? (Identify all stakeholders in the business including the customer)
<insert comments on the completion of the stakeholder analysis>
How will benefits be tracked? (Have they been adequately defined?)
<insert comments on benefits metrics>

Business change

Will the project change the way people do business? (Will people need to work differently?)
<consider if the project will change the way that 'normal business' is conducted>
Is the business ready for the project? (Are training needs identified or other organizational changes needed?)
<consider what else is being done in other parts of the business related to the project>

Scope definition

Has the scope been defined? (What level of feasibility work has been done?)
<insert comments on the accuracy of the scope of the project>
Have the benefit enablers been defined? (Will the project enable the benefits to be delivered when the project is complete?)
<insert comments on how the scope is linked to the business benefits>
Have all alternatives been investigated? (Which may include *not* needing the project)
<insert comments on all alternatives to this project which have been considered>
Have the project success criteria been defined and prioritized?
<consider the areas of scope which the project requires to be completed in order to deliver the business benefits>

Stage One decision

Should the project be progressed further? (Is the business case robust enough for detailed planning to commence?)
<insert the decision — yes or no — with comments>

Table 3-2 Customer versus sponsor

Project type	Sponsor role	Customer role
Engineering capital project	Either the Head of the Engineering Projects Group or the Site or Operations Director	The group who will eventually manage the capital asset
Manufacturing process improvement project	The Operations Director	The group who manage the current manufacturing process and will continue to do so after improvement
Product development project	Either a product champion in research and development or new product development or a strategic Manufacturing Director	The manufacturing organization who will be required to manufacture the new product *and* the marketing part of the organization responsible for that product type

However before significant work is progressed the following questions need to have been positively answered:

- Has a business case been developed?
- Have all benefits been identified? (Why is the project being done?)
- Who is the customer?
- How will benefits be tracked?
- What are the benefit metrics?

Additional tools to support the definition of the benefits which are contained in this chapter are the Benefits Hierarchy (see page 20); the Benefits Specification Table (see page 26) and the Business Case Tool (see page 32). These tools support the articulation of the benefits at an early stage — they can also be used as the project definition progresses.

Benefit metrics are those measures which will confirm, after the completion of the project, that the business is realizing the benefits — the reason why the project was done in the first place.

During this analysis other key stakeholders need to be identified, for example the customer. The customer is different from the sponsor. The customer is the eventual 'user'; whereas the sponsor may only have an overall accountability for the area in which the customer operates. Typical roles for different types of project are shown in Table 3-2.

Business change

Every project has the potential to change the nature of the business or organization into which it is being delivered. These checks therefore challenge whether enough is known about these changes:

- Will this project change the way people do business?
- Is the business ready for the project?

Traditionally projects have commenced with a request for a specific scope and ended when that scope has been delivered. In today's projects it has been recognized that projects need to be robustly connected to the organization. For example, there is no point in upgrading a manufacturing facility if the production operators are not retrained to operate the upgraded facility. The issue of training is a business change issue which is commonly omitted — it traditionally isn't a part of the scope of a 'hard' engineering project although it is commonly included in business change projects.

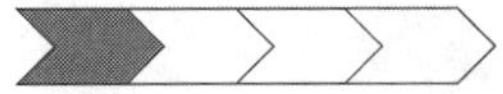

At a very early stage in a project it is critical that all aspects of business change as a result of a particular project are assessed. They can then be defined and managed either within the project scope or by the customer.

Scope definition

The project scope should enable the business benefits to be delivered. The project scope does not deliver the business benefits itself. This section of the checklist is therefore challenging whether the scope definition is robust enough:

- Has the scope been defined?
- Have the benefit enablers been defined?
- Have all alternatives been investigated?
- Have the project success criteria been defined and prioritized?

For many project and business managers the distinction between scope and benefits is the hardest thing to define at the early stages of a project:

- Project scope — what needs to be delivered in order for the benefits to be realized, sometimes referred to as the benefits enabler.
- Benefits criteria — the reason the project is being done; the articulation of the benefits to the organization.

Additionally, within the scope specific objectives can be defined; within the benefits criteria benefit metrics can be defined:

- Project objectives — measure scope delivery and can also be referred to as benefit enablers. Those objectives which measure areas of scope which are critical to the delivery of the project are called critical success factors (CSFs).
- Benefits metrics — measure the delivery of the benefits to the business; benefits delivery is often called benefits realization.

Stage One decision

Once all the available project information has been gathered it is critical that the Project Manager, in partnership with the project sponsor, makes a clear decision on the robustness of the work to date and asks:

- Should the project be progressed further?

The end point for Stage One is an agreement that the project has a robust business case in the context of the organization into which it will be delivered.

Using the 'Why?' Checklist

This tool can be used in many different ways:

- To support development of a robust business case.
- To check the progress of an idea under development.
- To audit a project where delivery has already commenced.

However, it has the most power when used in a team context — to support understanding of the idea, potential project or 'live' project. In this way, although a lot of hard data is required to complete the checklist, it can be done in such a way as to manage and support some of the softer project issues, for example the development of:

- A real understanding of the customer's needs.
- A true partnership with the project sponsor.
- A shared understanding of the project scope with the Project Team.

Each of the case studies in this chapter uses the 'Why?' Checklist to demonstrate different uses.

In completing the checklist it can be seen how the use of other tools within this chapter can support the development of the business case and therefore the completion of the Stage One decision.

Tool: Benefits Hierarchy

The aim of the Benefits Hierarchy is to support the completion of the detailed definition of the idea for the project and the development of a robust business case. The completion of this tool enables positive responses to the checks in the 'Why?' Checklist introduced in the previous section (see page 15).

The Benefits Hierarchy is a tool which confirms the alignment of the intended project scope to the targeted business benefits within a specific organization; it can identify additional benefits and also support the identification of high-level scope gaps, for example project CSFs.

The Benefits Hierarchy explained

The tool is made up of two distinct parts:

- The generation of an early Stage One Simple Benefits Hierarchy — used to develop an idea and to align that idea to an organizational benefit.
- The generation of a later Stage One Detailed Benefits Hierarchy — used as a working document to collate early project information and to continually assure alignment of the project to the business.

So although the Simple and Detailed Benefits Hierarchies are the same tool they are used at different times and for different purposes.

The Simple Benefits Hierarchy

This can be completed at a very early stage in the idea development. It is used to determine if a specific idea for a project would actually deliver a benefit contributing to the achievement of the organization goals or strategy — this is called a benefit criteria.

The template for this tool is simply a triangle containing five distinct levels as shown in Figure 3-3. The triangle asks for answers to five questions (one per level in the triangle) which assure all stakeholders that there is value in developing the idea further:

- *Benefit criteria* — why is the project being done?
- *Benefits business case* — what is the cost/benefit analysis?
- *Benefit enablers* — what does the project have to deliver to enable the benefits to be realized?
- *Project objectives and critical success factors* — what will be measured to prove the project has been delivered?
- *Benefits realization* — what will be measured to prove the benefits have been realized?

The Simple Benefits Hierarchy should be a succinct summary of the overall idea and forms a suitable 'starting point' for the later development of the Detailed Benefits Hierarchy.

The Detailed Benefits Hierarchy

This is usually completed at a later point in the development of the potential project once it has been shown to have a link to the delivery of an organizational benefit. This tool is a 'working document' and requires a more detailed review of each level given in the Simple Benefits Hierarchy. The template reflects this by converting the triangle to a table format. The Detailed Benefits Hierarchy requires more definition of the potential project.

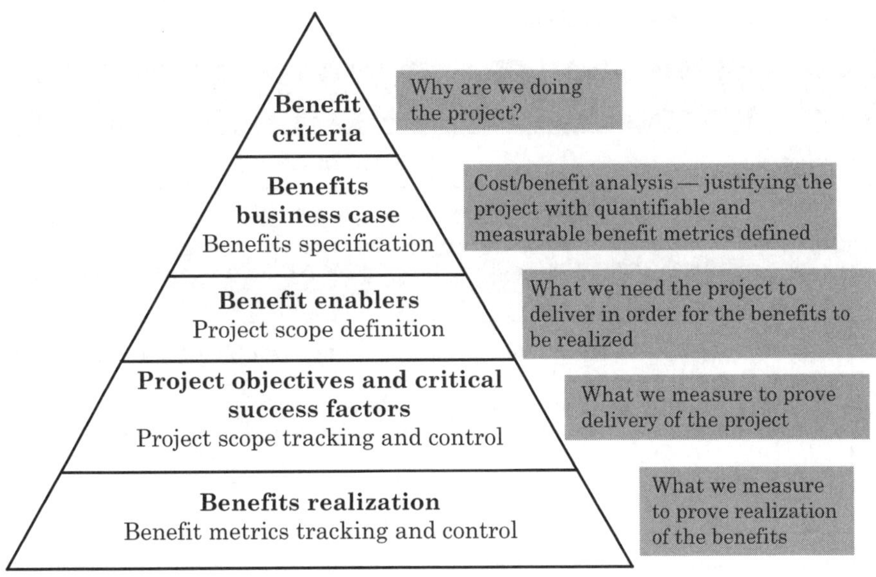

Figure 3-3 The Simple Benefits Hierarchy

In completing this level of detail, the scope of the potential project is challenged thus ensuring a robust link to the required benefits which are intended to be delivered. Table 3-3 shows the Detailed Benefits Hierarchy with high-level guidance on how to complete each section. The following are additional notes to support tool completion. Note that the completion of the other tools in this chapter will either support, or be supported by, the completion of the Detailed Benefits Hierarchy.

Benefit criteria

Typically the benefits criteria will be linked to an organizational objective which the project will support or enable. As most organizational objectives tend, by their very nature, to be strategic the use of a benefits mapping technique can support the translation of these high-level goals into benefit criteria.

The aim of defining benefit criteria is to support the ease of alignment of organizational activities, such as projects, to the strategic goals of the organization.

Business case

The Business Case Tool (see page 32) is completed using the information contained in the Detailed Benefits Hierarchy. The data for the Detailed Benefits Hierarchy is usually from a specific organizational procedure which requires all projects to be appropriately defined and a cost estimated.

Project scope

The 'Why?' Checklist (see page 15) supports the completion of this part of the hierarchy. It challenges whether the scope has been robustly defined and whether all suitable alternatives have been considered.

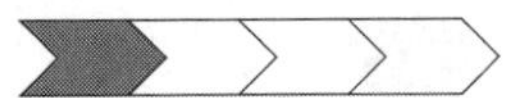

Table 3-3 The Detailed Benefits Hierarchy explained

Project Management Toolkit **— Detailed Benefits Hierarchy**		
Project: *<insert project title>*		**Date:** *<insert date>*
<insert brief project description here>		
Level in hierarchy	**Item in hierarchy**	**Comment**
1 Benefit criteria	*<insert the specific benefit criteria in here as shown in the Simple Benefits Hierarchy — use the full title of each criterion>*	*<confirm which of the benefits criteria are the major benefits and therefore will drive the project categorization>* *<note benefits criteria which may be enabled by the project or which may be delivered as a by-product of the project, that is to say they are not the main reason for doing the project>*
2 Business case	*<insert the business case as shown on the Simple Benefits Hierarchy with as much detail as has been developed>*	*<expand upon the business case so that the cost/benefit relationship is absolutely understood — the data will be used to generate the business case>*
3 Project scope	*<insert the agreed project title in here as this is the summary of the agreement of scope with the sponsor — the scope will be a part of the agreement with the sponsors and is what will have to be delivered if the project is approved>*	*<insert a comment to fully explain what the project is all about in plain English — It is usual for the project name to be challenged again as the scope is better defined. It is at this stage of the Detailed Benefits Hierarchy that deviations are seen from the original Simple Benefits Hierarchy. This is usual and driven by the idea development work which has been completed>*
4 Project objectives	*<insert the Level 1 CSFs here — as a part of the development of the idea the scope may have been further developed and the CSFs may have changed>*	*<ensure that the CSFs are valid — do they support realization of the benefits outlined in the business case?>*
5 Benefit metrics	*<insert specific metrics with targets as appropriate which are generated from the Benefits Specification Table>*	*<comment if the data to enable this metric to be tracked is routinely measured now and if not how/when this should commence — this may be a key issue when pulling together the business case>*

Project objectives

The development of a simple path of CSFs (see page 70) supports the completion of the Detailed Benefits Hierarchy.

Benefit metrics

The development of a Benefits Specification Table (see page 26) supports the completion of this part of the Detailed Benefits Hierarchy.

Using the Benefits Hierarchy

This tool has both 'soft' and 'hard' functions:

- It is a *stakeholder management tool* — it can support the generation of sponsor and team buy-in, it can assist with showing the need for change (through highlighting the degree of dissatisfaction with the current situation) and in building support for the change.
- It is a *team management tool* — it can support the generation of ideas and constructive discussion around the detail of the idea under development. It further allows team members to reflect on the value of the changes to them as someone within the business area (targets); as well as considering the overall impact on the service under review (to customers and the organization).
- It is a *scope and benefits definition tool* — the Simple Benefits Hierarchy is the start of defining what the idea would deliver as a project — the CSFs and the 'why' — the benefit metrics. The Detailed Benefits Hierarchy requires detailed and specific information about the project. This enables the business case to be generated, challenged, reviewed and updated until it is robust and able to be delivered by a defined activity, within a defined time and via management of measured risks.

The Simple Benefits Hierarchy

In order to complete the initial Simple Benefits Hierarchy template key stakeholders should be brought together in a 'facilitated brainstorm' where they should:

- Draw a triangle on a blank flip chart and mark the five sections (Figure 3-4).
- Start building the Simple Benefits Hierarchy from an 'idea' that in most cases will have already been developed into a draft scope or business case.
- Work upwards to determine if the benefit delivered aligns with the organizational goals/strategy:
  - Ensure that the idea is converted into an appropriate project title which accurately and briefly describes the project scope — the things that have to be delivered in order for the benefits to be realized.
  - Detail the benefits that will be realized, for example improvements in service, cost reduction, and improved customer satisfaction. These are all potential business cases for a project.
  - Once the business case is confirmed and matches the project scope the specific type of benefit which will be delivered should be defined. The major benefit identified will usually determine the project categorization.
- Work downwards to confirm the CSFs and then the Benefit Metrics:
  - In order to develop the CSFs an input–process–output diagram (Figure 1-2 on page 2) can be used as a precursor to the development of the critical path of CSFs — the output is the required scope and the CSFs are the process to achieve this (see page 70). Only the first level of CSF is needed (see page 70).
  - To develop appropriate benefit metrics the Benefits Specification Table can be used (see page 26). The team should try to define specific metrics which have value in demonstrating that the business case has been met.
- The team should review the completed triangle on the flip chart and, if all agree that the idea is valid and the triangle fully aligned, it should be transferred to the formal Simple Benefits Hierarchy template for onward communication and use.

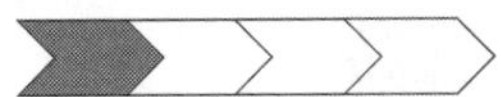

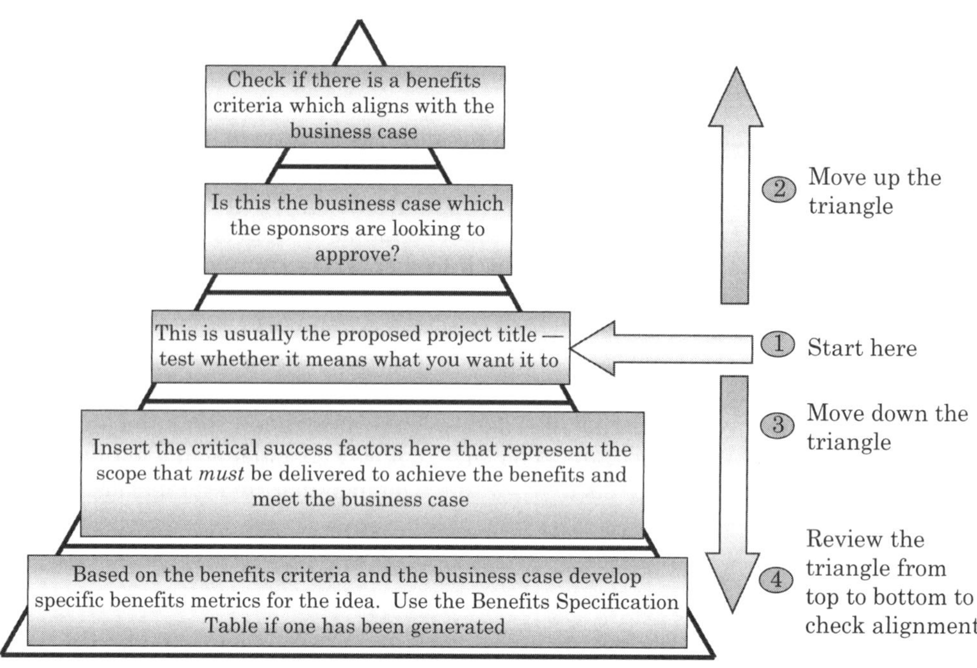

Figure 3-4 Completing the Simple Benefits Hierarchy

The completed Simple Benefits Hierarchy can then be used by stakeholders to 'test' the idea and to continue to build support for the change within the business area and within the leadership of that business area — targets, sponsors and reinforcing sponsors.

This simple visual representation of the potential project or idea can be used to great effect to support communication and to generate further support and understanding. An example of a completed Simple Benefits Hierarchy is shown in case study C (see page 45).

The Detailed Benefits Hierarchy

The Detailed Benefits Hierarchy is a team tool managed by the Project Manager. It can be generated if the team have an agreed, succinct, Simple Benefits Hierarchy and appropriate team working processes.

For example, the Detailed Benefits Hierarchy can be introduced at Project Team meetings to facilitate an open session on the reason the project is being done (challenge the 'why'). Note that at this stage new team members may join the project and it is critical that they 'buy-in' to the Simple Benefits Hierarchy.

As the idea is further defined through individual and team activities (including sponsor input) the hierarchy table template should be completed (ideally it is kept on a shared area for ongoing review by the team). The team should develop the 'what', 'when' and 'how' to measure benefits delivery.

The Detailed Benefits Hierarchy is in effect a working project document which enables the succinct business case to be formulated (see the Business Case Tool, on page 32) using a portfolio of tools.

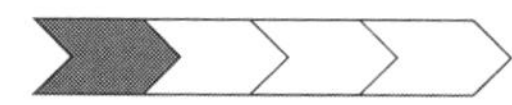

The completed Detailed Benefits Hierarchy then becomes a valuable tool for further stages of the project following authorization, as well as a key internal Project Team tool providing:

- *Baseline data* — the hierarchy sets the baseline in terms of what the project must deliver to meet the commitments as authorized via the business case.
- *Consistent communication* — the hierarchy can be used to ensure consistent communication to all new stakeholders who may get involved once the project is approved and moves into subsequent stages. For example, new Project Team members and a wider customer group.

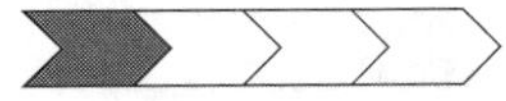

Tool: Benefits Specification Table

The aim of this tool is to support the identification of the potential benefits from a project. Benefits can be categorized in many ways however the simplest is to consider our ability to quantify and value them:

- *Financial benefits* — these are 'hard' benefits which most projects aim to deliver, for example cost savings, capital cost avoidance, and reduced inventory. Financial benefits are very tangible to an organization and usually very measurable.
- *Non-financial benefits* — these are 'soft' benefits which typically business change projects would deliver, for example increased employee motivation and improved customer satisfaction. Non-financial benefits are less tangible to an organization and are also harder to measure.

Traditionally organizations have focused on the financial benefits which will be realized by a project, particularly if it involves an engineering capital project. This has been due to the ability to measure them and also the focus on financial payback on investments within organizations.

Today it is recognized that less tangible benefits can be measured and are as valuable to an organization. This tool, therefore, aims to support the identification and articulation of all appropriate benefits which a specific project would enable.

The Benefits Specification Table explained

The table has six columns:

- Potential benefit.
- Benefit metric.
- Benefit metric baseline.
- Accountability.
- Benefit metric target.
- Area of activity.

Table 3-4 shows the Benefits Specification Table with high-level guidance on how to complete each column. The following are additional, more detailed, notes to support tool completion.

Potential benefit

These are the high-level benefit criteria which link to the achievement of the organizational objectives. The benefit metrics link directly to these criteria. For example this might be operational effectiveness if the project was a revamp of an existing manufacturing facility. In order to develop the appropriate relationship between organization objectives and benefit metrics it is usual to build a benefits map (Figure 3-5).

Benefit metric

These are the measurable metrics which, if delivered, support the organization in meeting their organizational objectives. It is critical that appropriate units are indicated. For example this could be the increase in equipment utilization if the project was an equipment replacement programme then an appropriate unit would be a percentage.

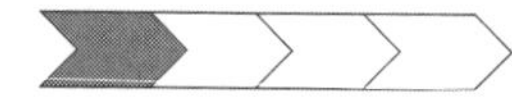

Table 3-4 The Benefits Specification Table explained

Project Management Toolkit — Benefits Specification Table					
Project: *<insert project title>*			**Date:** *<insert date>*		
Potential benefit	**Benefit metric**	**Benefit metric baseline**	**Accountability**	**Benefit metric target**	**Area of activity**
What the project will enable the business to deliver	Characteristic to be measured	Current level of performance	Person accountable for delivery of the benefit to target	Required performance to achieve overall benefits	The project scope that will enable this benefit to be delivered
<insert the benefit criteria which this benefit metric is related to>	*<insert the specific benefit metric which is to be measured>*	*<insert the measured level of current performance, that is 'before' project commencement>*	*<insert the name of the person who is accountable for the delivery of this specific benefit>*	*<insert the target level of the metric at the target date>*	*<insert the area of the project which is specifically related to the delivery of this benefit>*

Benefit metric baseline

This is the level of the benefit metric prior to the delivery of the project, sub-project or activity (project scope). Where baseline levels are unknown then some initial data collection is needed if the business case is to be robust. It is critical that the units are consistent during all baseline and subsequent measurement. Any calculation must be clearly stated.

Benefit metric target

This is the end-point for the benefit metric that is aimed for and links directly to the achievement of the stated business case. Some benefits are realized during the delivery of the project scope whereas others will trend towards the target only after the completion of the project. However each benefit metric should have a forecast level against the schedule. For some benefit metrics it is appropriate to monitor the trend closely; for others a review at specific points is all that is necessary.

Accountability

This is the person who is accountable for the delivery of the benefit. It is usual for this to be a member of the customer team — the team who will accept the completed project. A key part of this role is the development of how the benefit metric can be reliably measured and then to develop the baseline.

Activity

This is the part of the project scope which will contribute towards, or completely deliver, the increase in the benefit from the baseline level to the target level. This may be a specific CSF or an entire sub-project. For example the installation and commissioning of new equipment could directly contribute to an increase in equipment utilization.

Using the Benefits Specification Table

The basis for the generation of the Benefits Specification Table is benefits mapping. This is a method which identifies and articulates the benefits that relate to a specific organizational goal. The outcome of a benefits mapping session is directly input to the Benefits Specification Table.

Benefits mapping is an activity which relates organizational or area objectives to specific benefit metrics which would be delivered by specific activities, projects, sub-projects or CSFs.

A benefits mapping session is best conducted with the involvement of a cross-section of the organization who will be responsible for the achievement of the high-level goals from which the map starts. Figure 3-5 shows a generic benefits map.

In order to generate a benefits map key stakeholders should be brought together in a 'facilitated brainstorm' where they should:

- Write the 'organizational goal' at the top of a blank flip chart (Figure 3-5). If the benefits map is for a specific part of the organization then the benefits mapping can start from that business area's strategic goal (which aligns with the organizational goal).
- The group should be asked 'how will we achieve this goal?' and should write their responses on Post-it® notes and do the following:
  - The group are likely to put the objectives on the Post-it® notes which should be placed beneath the appropriate goal.
  - Group any similar objectives.
  - Don't throw away any Post-it® note. It may highlight an objective which may not be currently stated within the organization or business area.
  - Challenge each 'level two' item on the benefits map by saying 'if I do <objective> I will achieve <goal>' — if the causal link isn't there then put the Post-it® note to one side — it may actually be a 'level three' item on the benefits map.

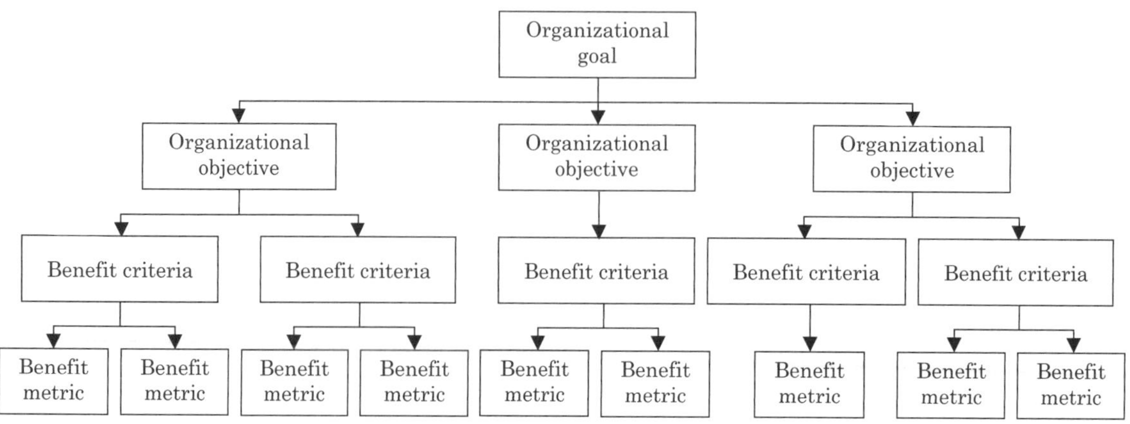

Figure 3-5 A generic benefits map

- The group should be asked 'how will we achieve these objectives?' and should write their responses on Post-it® notes and do the following:
  - The group are likely to put generic criteria on the Post-it® notes which should be placed beneath the appropriate objective.
  - Group any similar criteria.
  - Don't throw away any Post-it® note. It may highlight criteria which cannot be linked to an objective.
  - Challenge each 'level three' item on the benefits map by saying 'if I do <criterion> I will achieve <objective>' — if the causal link isn't there then put the Post-it® note to one side — it may actually be a 'level four' item on the benefits map.
  - Review all Post-it® notes left to one side from the previous level.
- The group should be asked 'how will we measure these criteria?' and should write their responses on Post-it® notes and do the following:
  - The group are likely to put generic metrics on Post-it® notes, which should be placed beneath the appropriate criterion.
  - Group any similar criteria.
  - Don't throw away Post-it® notes that highlight criteria which cannot be linked to an objective.
  - Challenge each 'level four' item on the benefits map by saying 'if I measure <metric> I will prove achievement of <criterion>' — if the causal link isn't there then put the Post-it® note to one side.
  - Review all Post-it® notes left to one side from the previous level.
- The group should be asked to brainstorm all the things they would measure to prove that they are achieving their organizational goal:
  - The group are likely to think of current projects and activities.
  - These benefit metrics should be compared to those on the benefits map — any additions?
- Start at the bottom of the benefits map and check the links to the goal.

The mapping highlights the benefit criteria which are the high-level benefit categories and also the specific benefit metrics within these categories. In this way the benefits can be defined in a very specific manner; this tends to be of most use to the project cost/benefit analysis.

The Benefits Specification Table can then be populated and used in further discussions with the sponsor to confirm understanding prior to and during the development of the business case. Although a generic Benefits Specification Table can be generated from the benefits map the actual table for a specific project should be very specific to that project.

Figure 3-6 is an example of a benefits map and Table 3-5 is the Benefits Specification Table generated from the map. The project from which these were taken was a cleaning improvement project on a manufacturing site. Cleaning appeared to have spiralled out of control in terms of both cost and quality leading to a culture of 'there's always someone to clean up after me'. It was starting to become both a safety and a quality issue. The benefits map was used to understand the key criteria which aligned with the organization goal — a clean site at a benchmark cost — where that benchmark is appropriate for a site of that type.

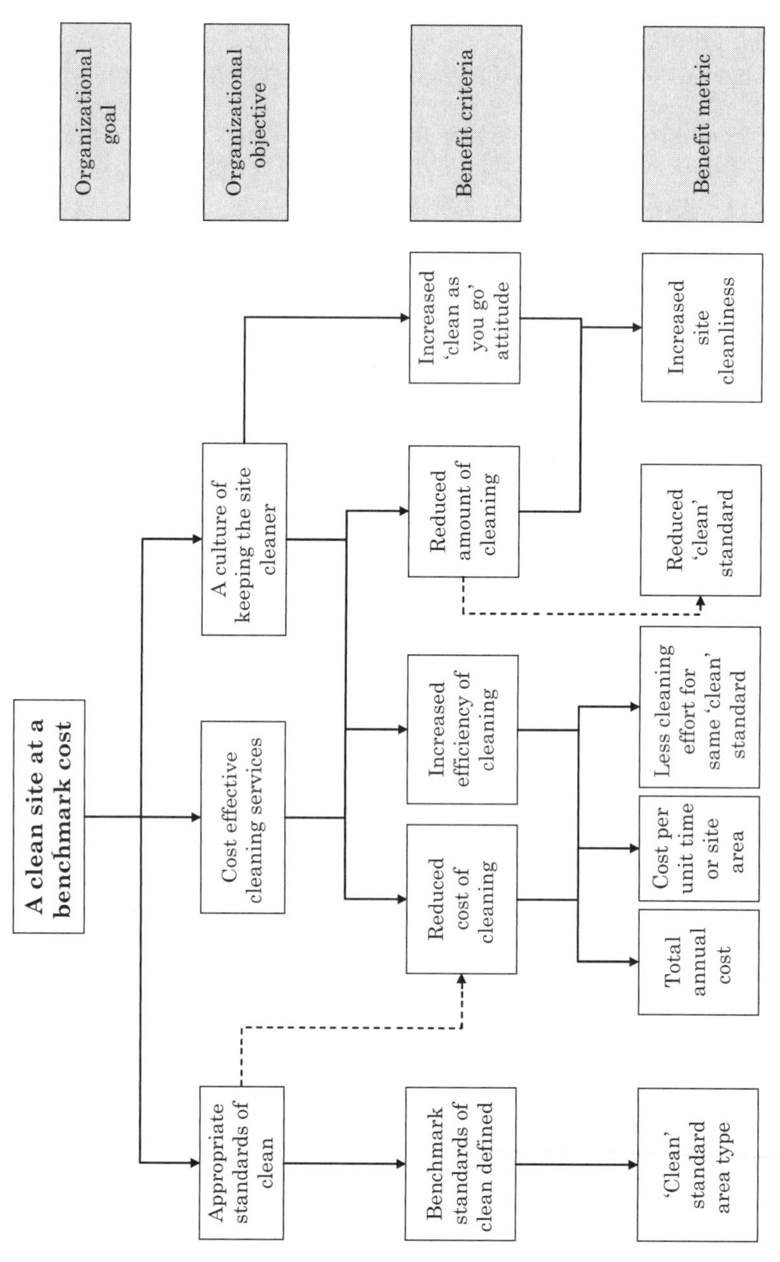

Figure 3-6 An example benefits map

Table 3-5 An example Benefits Specification Table

***Project Management Toolkit* — Benefits Specification Table**

Project: *<insert project title>* **Date:** *<insert date>*

Potential benefit	Benefit metric	Benefit metric baseline	Accountability	Benefit metric target	Area of activity
(What the project will enable the business to deliver)	*(Characteristic to be measured)*	*(Current level of performance)*	*(Person who is accountable for delivery of the benefit to target)*	*(Required performance to achieve overall benefits)*	*(The project scope that will enable this benefit to be delivered)*
Reduce the cost of cleaning	Annual cost of cleaning	£*x* per annum	Site Facilities Manager	50% of baseline	Facility clean-up project
	Hourly rate for cleaning services	£*x* per hour	Procurement Manager	80% of baseline	Cleaning contractor re-tender project
Reduce the amount of cleaning required	Measure of the 'level of clean' by area before cleaning	Not currently measured — baseline to be established	Site Facilities Manager	Target to be determined after baseline activity complete	Facility clean-up project
Increase the efficiency and effectiveness of cleaning	Measure of the 'level of clean' by area after cleaning	Not currently measured — baseline to be established	Procurement Manager	Target to be determined after benchmark activity complete	Cleaning contractor retender project

Selected benefits criteria were transferred to the Benefits Specification Table (Table 3-5) and two linked improvement projects were developed:

- *Facility clean-up project* — a project to reduce the cleaning costs per annum through linking to the cleaning re-tendering project (the right cleaning resource) and through the development of a more appropriate site 'clean' culture (which can impact more than cost).
- *Cleaning re-tendering project* — a project to redefine the type of contractor needed and the work they should be doing at an appropriate benchmark cost for the type of cleaning.

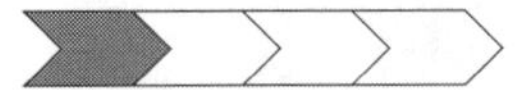

Tool: Business Case Tool (value-add or not?)

The aim of this tool is to succinctly articulate the business case for a potential project or idea thus enabling authorizing sponsors to approve or otherwise. If a robust business case cannot be generated so that the 'value-add' is clear then it is highly likely that there is not a robust business case and further development work is required.

If further work is completed and a robust business case can still not be found then it is clear that that the project should not be progressed any further.

The Business Case Tool explained

This tool is a one-page summary which conveys the 'value-add' for the organization in progressing the project further. The tool has seven main sections:

- Business background.
- Project description.
- Delivery analysis.
- Business change analysis.
- Value-add analysis.
- Impact of not doing the project.
- Project approval.

Table 3-6 shows the Business Case Tool with high-level guidance on how to complete each section. The following are additional, more detailed, notes to support tool completion.

Business background

The crucial part of this section is the articulation of the 'problem statement'. This is the problem which the project is proposing to solve for the organization and should have been generated through root cause analysis of the initial issue which was highlighted.

One root cause analysis technique is to conduct a 'five whys' exercise on an issue — if you ask 'why?' five times (no more, no less) then you will identify the root cause of a problem rather than the symptoms. Figure 3-7 shows an example based on a site cleaning issue — 'why is cleaning costing us so much?'.

The problem statement from this example is the root cause: There is no 'clean' culture on the site. It is this that the project scope must address.

This is a very simple example, a root cause analysis would usually end up as a complex map of diverse symptoms with the potential for a number of root causes as each route is followed through the map.

Project description

This should be a succinct description of the project although it may be useful to attach a Simple Benefits Hierarchy (see page 20) as this is an effective communication tool linking the project back to the organizational goals.

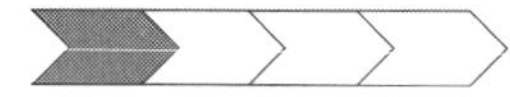

Table 3-6 The Business Case Tool explained

Project Management Toolkit — Business Case Tool			
Project: *<insert project title>*		**Date:** *<insert date>*	
Business case developed by	*<insert the name of the Project Manager developing the business case>*	**Date**	*<insert the date issued for approval>*
Project reference number	*<insert the project reference number as it will be referred to within the organization>*	**Business area**	*<insert the business area or organizational group who initiated this project>*
Project Manager	*<insert the name of the person who becomes accountable for the delivery of the CSFs if the project is approved>*	**Project sponsor**	*<insert the name of the senior organizational stakeholder who becomes accountable for the delivery of the benefits from the project if it is approved>*
Business background	*<insert a description of how the potential project was identified, within which area of the business and a summary of the problem statement>*		
Project description	*<insert a brief description of the project — the scope and the CSFs. If necessary attach the Simple Benefits Hierarchy which will have been updated during Detailed Benefits Hierarchy development (do not attach this as it is an internal project tool)>*		
Delivery analysis	*<insert a summary of the resources required to deliver the project (internal and external), capital and revenue; insert a summary of a high-level risk assessment and include any dependency issues related to other 'live' projects or activities within the organization, attach a preliminary milestone schedule>*		
Business change analysis	*<insert comments regarding the potential impact of the project should it be approved — the impact of this project on the business>*		
Value-add analysis	*<insert the summary of the Detailed Benefits Hierarchy in terms of the high-level benefit criteria, the cost/benefit analysis and the specific benefit metrics; discuss any link to key milestones>*		
Impact of not doing the project	*<discuss the level of urgency regarding the project — what happens if the project is delayed or is a lower priority than another? Increased costs? Customer dissatisfaction? Lost customer orders? What is the urgency of the problem which the project will solve?>*		
Project approved *(Value-add or not?)*	Yes/No *<approval to proceed to the next project stage gate>*	**Name of approver and date**	*<insert the name and position of the person approving the project, signed and dated by them>*

Delivery analysis

Up to this stage in a project the focus has been on justifying the project's existence and little formal delivery planning will have been done.

In order to complete this section it would be usual to consider the high-level planning issues which are contained in the 'How?' Checklist (see page 56).

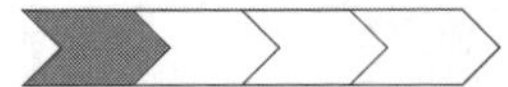

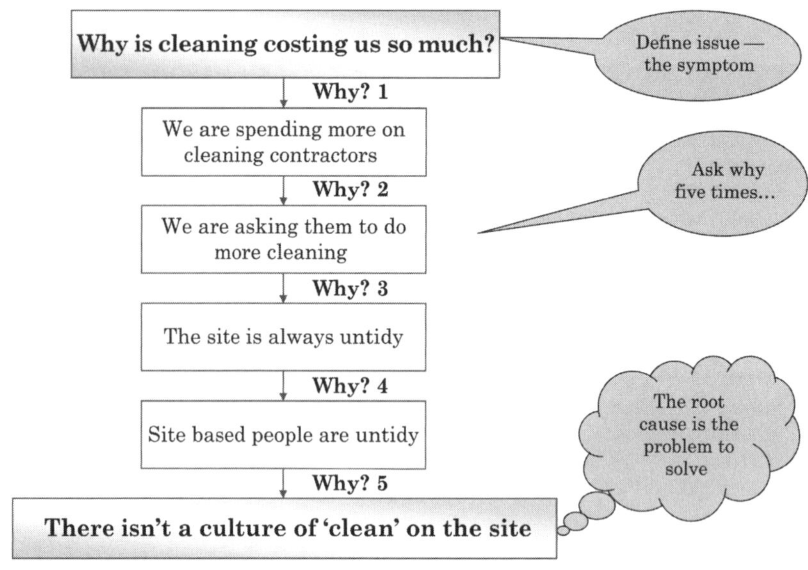

Figure 3-7 An example five whys analysis

In terms of formal planning it is usual to give an indication of the resources required to complete the next project stage (Stage Two — Project Delivery Planning) and an approximate time-line. For more traditional projects (such as engineering build projects) preliminary cost and schedule planning can be completed based on historic benchmarks.

Business change analysis

This section can be completed using the data collected during the use of the 'Why?' Checklist (see page 15). It is critical that a succinct statement is made regarding:

- The potential impact of the project on the business.
- How the business may need to change if the project is to be successful.
- Related activities which are out of the control or influence of the Project Manager and in some cases the sponsor.

In this way additional support can be requested and formal links to other projects and business activities made.

Value-add analysis

For capital projects this section is easily completed using the usual costing techniques versus the data compiled in the Benefits Hierarchies:

- Factorial cost estimation based on the main items to be purchased and comparison with historic data trend for the entire project cost and time.
- Traditional benefit criteria linked to security of supply to the marketplace (for example more capacity and security of compliance) and/or decreased operational costs (for example decreased cost to make and supply a product).

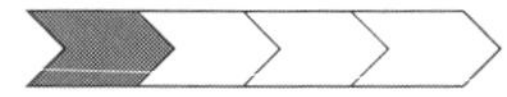

For non-capital projects early stage cost estimation is rarely completed to the same degree for a number of reasons:

- The problem being solved is 'softer' and fairly unique so that historic norms cannot be used to assess the likely revenue costs or the overall timeline for implementation.
- The cost/benefit analysis is focused on proving a benefit link.

The level of value analysis is likely to impact on the type of approval which an organization will give (see project approval section on page 36), however in order to actually demonstrate that the project will deliver value to an organization there must be a robust cost/benefit evaluation prior to delivery.

Impact of not doing the project

The basis for this section would usually be a 'what-if' scenario analysis. This is a risk assessment technique which builds scenarios against a 'what-if' statement.

The analysis looks at the worst-case scenario and then steps back to look at the scenarios leading to this. For the example in Figure 3-8 the worst-case scenario is that the company loses market share:

- If the new plant isn't built then there will need to be investment in the current plant to stop the occurrence of the worst-case scenario.

The analysis also demonstrates that, in the example, any plant improvement option is only delaying the inevitable investment in a new build, as the product forecasts are so healthy and long-term.

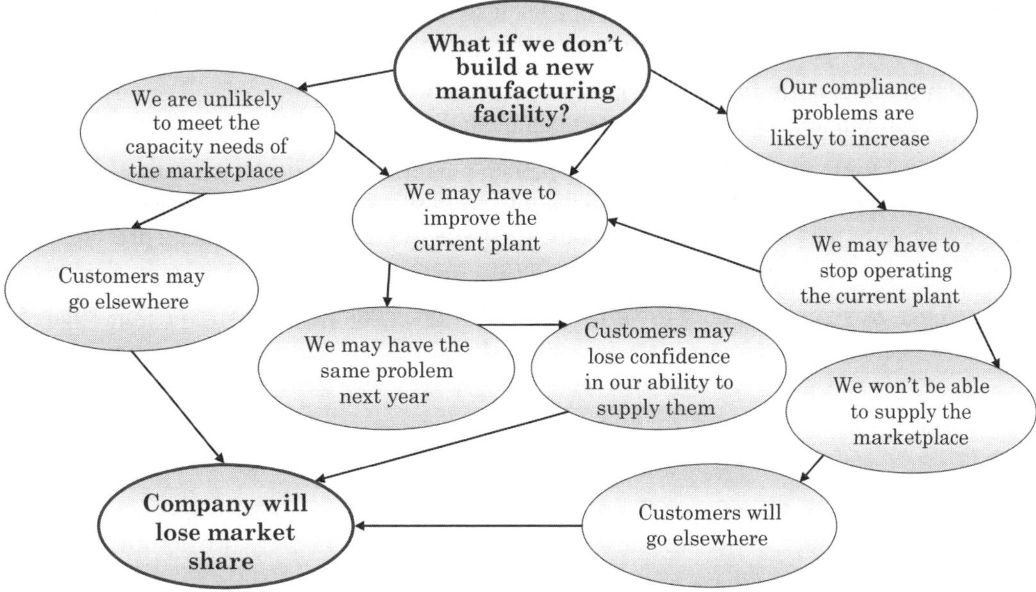

Figure 3-8 'What-if' scenario example

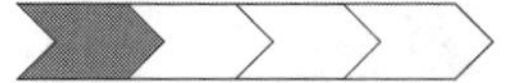

A good test of the reliability of the 'what-if' technique is whether you believe that the scenarios are possible. If you don't believe they are you won't be able to convince others and it may well be that the impact of not doing the project is minor.

Project approval

The definition of what the specific signature actually means — what action it drives — must be agreed within individual organizations. However, the successful sign-off must also agree that it is 'value-add' for the project to proceed in terms of: the delivery of business benefit versus the expenditure of business funds (capital, revenue and resources) versus the impact of not doing the project.

In many organizations the procedures for approving capital funding are well established and based around complex investment payback calculations — this tool does not replace these procedures but adds another dimension in the approval process — the articulation of the 'value-add' in terms other than purely financial.

For revenue-based projects the approval process for funds is typically done annually and budgets are agreed and devolved into the organization for the following year. For this case the tool can support budget holders in choosing the highest priority projects based on the 'value-add' they would bring to the organization.

Additionally the tool highlights the need to formally identify both the sponsor and the Project Manager; the Project Manager for Stage Two onwards is not necessarily the person who has developed this business case.

Using the Business Case Tool

As indicated in the project approval section on above, the tool can be populated from a number of sources, depending on current organizational procedures for approving ideas or a potential project. The Detailed Benefits Hierarchy could be used as one such source. The completed Business Case Tool should be able to support:

- Requests for funding to do further feasibility or costing work.
- Requests for full funding so that the project can be delivered.

It would be usual for the Project Manager to complete the tool and to attach relevant data to support a discussion with the sponsor and then for the sponsor to request support from the business.

The most important part of generating a business case is the ability to succinctly present the 'value-add' to any appropriate audience; this is also the hardest part:

- If you can't say why you are doing the project in a few lines you probably haven't articulated this well enough.

Case study A — if only the 'Why?' Checklist had been around then!

This case study is based on a project which when completed did not deliver the benefits for which it had been approved. The 'Why?' Checklist is used to trouble shoot — 'why not?'

Improving production efficiency

Situation

During a normal month-end review a production manager noticed that over the last couple of months the maintenance and operation costs had started to increase. The plant was over ten years old and was fairly manual requiring a team of four operators (three shifts covering 24 hours) and on-call maintenance team support. It was clear that the plant now required an increased operations and maintenance personnel presence to keep the plant running.

The production manager had been thinking for a while that the plant needed to utilize newer technology and firmly believed that increased automation would deliver efficiency savings. He therefore completed a request for capital funding (automation of manufacturing plant) and submitted this to the finance manager. As the funding was within the site authorization limits it was approved immediately with little back-up documentation but with an assurance that the increasing plant operation costs would be controlled within budget and with the possibility that overall costs would be decreased as a smaller operations team would be needed in the future.

The project

A specialist consultant reviewed the manufacturing plant and a new distributed computer controlled system (DCS) was selected as the appropriate design solution. This would allow the operators to be based in the upgraded control room and to operate the plant remotely.

The design of the software took longer than anticipated due to the complexities of the operation and the criticality of certain process steps; some steps had to be left in manual operation for this reason with the DCS merely acting as a data logger.

The commissioning was extended due to the operator training issues with the mix of manual, semi-automatic and automatic operation.

The outcome

Six months after the project was completed the operations manager was reviewing his monthly operating costs with the site manager:

- Overall operating costs had actually increased due to the revised preventative requirements of the automated plant, although operator overtime had reduced.
- The total number of operators running the plant remained the same.
- The payback for the capital investment could only be measured in increased reliability of the plant. Once the post commissioning problems had been solved the plant was back to running at the pre-project throughput.

The site director asked for a formal review (Table 3-7) so that he could understand:

- Whether the increased reliability could be measured in terms of the reduced emergency maintenance call-out? — Apparently this wasn't measured.
- Why the total numbers of operators using the plant had not reduced? — The complex nature of the plant still required all operators for a key processing step.
- Why the plant wasn't producing at a higher throughput? — This wasn't the point of the project.

The operations manager and site director seemed in disagreement about whether the project had been a success. There was no doubt that the operators were happy with the revamped plant:

- Fewer breakdown 'hassles' — previous to the project each day started with a 'find the fault' session.
- More automation for the simple processing steps (represented 60% of the process) — up-skilling them and reducing some of the high manual workload.

Table 3-7 A completed 'Why?' Checklist — case study A

Project Management Toolkit — 'Why?' Checklist

Project: Plant automation	**Project Manager:** Confidential
Date: Confidential	**Page:** 1 of 2

Sponsorship

Who is the sponsor? (The person who is accountable for the delivery of the business benefits)
The Operations Manager
Has the sponsor developed an external communication plan? (How the sponsor will communicate with all stakeholders in the business)
The Operations Manager communicated with the Finance Manager and the Site Manager when he requested funding approval (a sum of money). No other communication was deemed necessary

Business benefits

Has a business case been developed?
No. The Operations Manager assumed that some arbitrary capital amount was appropriate to spend on this facility on the basis of undefined efficiency improvements
Have all benefits been identified? (Why is the project being done?)
No calculation of return on investment was completed. The less tangible benefits (increased reliability) were not fully articulated — in fact this was almost an unintended benefit of the project. The assumption that fewer operators would be required was never discussed with the operations teams. Due to project changes the capital spent was 15% over that requested. No calculation was made of the revenue implications (lost revenue at plant shutdown)
Who is the customer? (Identify all stakeholders in the business including the customer)
The operations management team and the team of operators were not identified as stakeholders. The project was given to a Project Engineer who employed a Consultant and who then contracted with a larger organization. The Operators were only involved when the software was being developed and it became clear that the design needed 'tweaking'
How will benefits be tracked? (Have they been adequately defined?)
Benefits were not tracked — efficiency benefits would be seen in some of the monthly costs which are monitored

Business change

Will the project change the way people do business? (Will people need to work differently?)
The Operators will have to work quite differently as well as take on additional preventative maintenance activities (as they won't be as busy in a remotely operated plant). Because a part of the process has to have significant manual intervention the mix of automatic, semi-automatic and manual modes of operation were confusing for some Operators and frustrating for others
Is the business ready for the project? (Are training needs identified or other organizational changes needed?)
The Operators did need substantial training and this was not allowed for in the capital budget and had to be paid for out of the site training budget. The plant shutdown for the installation and testing works was longer than anticipated and some external customer orders could not be satisfied from the stock

(continued)

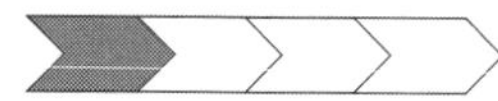

Table 3-7 (Continued)

<table>
<tr><th colspan="2">Project Management Toolkit — 'Why?' Checklist</th></tr>
<tr><td>Project: Plant automation</td><td>Project Manager: Confidential</td></tr>
<tr><td>Date: Confidential</td><td>Page: 2 of 2</td></tr>
<tr><td colspan="2">Scope definition</td></tr>
<tr><td colspan="2">Has the scope been defined? (What level of feasibility work has been done?)
No feasibility work was completed. The scope was determined by the Consultant based on an instruction to 'automate the plant'. The Operators were able to provide valuable plant information about which parts of the plant had to remain manual, for example, but this was only identified late in the design when the Consultant requested their input
Have the benefit enablers been defined? (Will the project enable the benefits to be delivered when the project is complete?)
The scope was not linked to any real benefits
Have all alternatives been investigated? (Which may include not needing the project)
No alternatives were reviewed
Have the project success criteria been defined and prioritized?
The project was deemed successful if the plant was automated</td></tr>
<tr><td colspan="2">Stage One decision</td></tr>
<tr><td colspan="2">Should the project be progressed further? (Is the business case robust enough for detailed planning to commence?)
This project did not have the required level of definition to proceed to delivery. The fact that the assumed benefits of the project were not delivered is no surprise (nor that the Site Director and the Operations Manager had different expectations). The scope was not linked to any organizational benefit</td></tr>
</table>

Case study B — using the 'Why?' Checklist to stop a project

This case study demonstrates what can happen when you use the 'Why?' Checklist on a project which appears to deliver little or no business benefit. In this instance the approved project was stopped.

Dryer room upgrade

Situation

Within a pharmaceutical manufacturing site there was a series of multi-product finishing rooms within a clean room suite (an area where the air quality, temperature and humidity were maintained within specified limits). One of these rooms contained a dryer. The dryer was loaded with a liquid product and the output, dried product, was immediately put into kegs and then dispatched from site (following confirmation of test results). This was a small-scale development operation with many changes of products in a working week.

The majority of the product was both highly active and highly toxic in some way and the operator and environment exposure limits had to be strictly controlled within set parameters.

As the dryer has to handle many different pharmaceutical products a critical issue is the cleanability of the room and the dryer between different products. Cross contamination of one product with another is not acceptable.

The dryer could not be cleaned *in situ* and so additional risks were being taken by having to remove the dryer from the room whilst 'dirty' and then cleaning in a separate wash area before being returned 'clean' to the cleaned dryer room.

To mitigate the additional risks (to the product, the operator and the environment) extensive procedures were in place which lengthened the product changeover time to such an extent that it exceeded processing time for many of the products. The cleaning was effectively driving the capacity of the dryer room and drying was becoming the bottleneck for the whole product development process.

It was decided that the dryer room needed to be upgraded.

The project

A young project engineer was brought in to deliver what was considered a relatively simple project. As the project engineer was unfamiliar with the good manufacturing practices (GMP) associated with pharmaceutical production operations the quality department gave him a GMP induction. The engineer was left in no doubt that compliance was his highest priority.

He proceeded to develop a plan and started to work with local site-based contractors who had been involved in other product room upgrades. He knew that in order to be able to open and clean the dryer in the room, the quality of the room environment had to achieve higher standards.

The preliminary front-end design work was completed and a ±10% cost estimate pulled together with an associated project schedule.

The project only needed final authorization to proceed and all the team expected that this was merely a formality.

The outcome

The authorization was declined and the project was stopped. A review (Table 3-8) ascertained that:

- The room was not big enough for the dryer to be opened up whilst *in situ* due to the opening mechanism type (the open dryer took up more than twice the floor space of the closed dryer).

Due to the criticality of some of the issues raised during the 'challenge' to the original project a new Project Team were put together and charged with a project to increase the overall capacity of the drying unit operation as this would directly improve the overall product development capacity of the plant (as drying is the bottleneck). A new product room was eventually constructed and the dryer installed with additional cleaning-in-place technology.

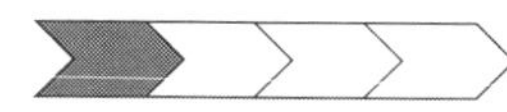

Table 3-8 A completed 'Why?' Checklist — case study B

Project Management Toolkit — 'Why?' Checklist

Project: Dryer room upgrade	**Project Manager:** Confidential
Date: Confidential	**Page:** 1 of 1

Sponsorship

Who is the sponsor? (The person who is accountable for the delivery of the business benefits?)
The Operations Manager although he appears unaware of this. The Engineering Manager appears to be pushing the project
Has the sponsor developed an external communication plan? (How the sponsor will communicate with all stakeholders in the business)
No — see above — not many people seem to know much about this project

Business benefits

Has a business case been developed?
The request for funding which was developed asks for a specific amount of capital with a claimed payback of 2 years based on the use of the room in processing more products than currently
Have all benefits been identified? (Why is the project being done?)
The capital request makes no mention of the cleaning problems which apparently initiated this project
Who is the customer? (Identify all stakeholders in the business including the customer)
The Production Manager
How will benefits be tracked? (Have they been adequately defined?)
This is not clear within the current project documentation

Business change

Will the project change the way people do business? (Will people need to work differently?)
It is intended to reduce a lot of manual movement of the dryer and change the overall cleaning philosophy between product changes
Is the business ready for the project? (Are training needs identified or other organizational changes needed?)
Operators are completely unaware of this project — when it was mentioned they highlighted that there was no way the dryer could ever be opened in its existing room. There isn't enough space

Scope definition

Has the scope been defined? (What level of feasibility work has been done?)
Yes but it now appears that this isn't the right scope for the problem which is supposed to have been solved
Have the benefit enablers been defined? (Will the project enable the benefits to be delivered when the project is complete?)
No the scope will not enable the dryer to be cleaned *in situ*
Have all alternatives been investigated? (Which may include not needing the project)
No
Have the project success criteria been defined and prioritized?
The current scope defines compliance and room environment as the critical issues for the project

Stage One decision

Should the project be progressed further? (Is the business case robust enough for detailed planning to commence?)
No — the project will not actually allow the dryer to be cleaned *in situ* — the reason for the project in the first place. The project needs to be stopped immediately and the problem reviewed again to look at other options

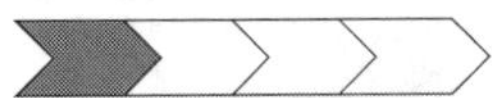

Case study C — using the why tools to justify a project

This case study uses all Four Stage One tools and is based on a real example in industry. The feedback after the project had been approved for delivery was 'this was a well defined project with a clear and sound business case'.

Bulk materials handling

Situation

An existing powder raw material, used in the production of a manufacturing site's largest product in terms of capacity and turnover, was identified by the UK Health and Safety Executive (HSE) as potentially hazardous to health as long-term exposure can cause breathing difficulties. Operational exposure limits have been set for the material which the company is obliged to comply with. The material is currently handled in bulk bags, which are dusty during and after use. Whilst the use of bulk bags is labour intensive for both production operators and warehouse staff the equipment required for the use of bags is minimal and this has been the driver for their current use.

The plant operations manager has investigated various options for equipment which will allow achievement of the exposure limits. During this development the option to install a bulk silo operation has shown an added benefit of a reduction in headcount, which would allow the project to have a financial justification as well as reducing operator exposure. Procurement have confirmed that the material is available in bulk from present suppliers and will investigate other alternatives.

There is no provision for the project in current financial plans so substitution against other identified capital funds will be required. However, the safety, health and environmental issues relating to the project are sufficient to allow conceptual development of the project and business case.

The project

The stakeholders for the project have been identified including off-site vendors and suppliers. The Project Team include project engineering, production, procurement, quality control and materials management as the project will alter the whole supply chain for the raw material into the production operation. External contacts have included raw material suppliers to review delivery options and also external powder handling companies to look at delivery forms they could provide. The site project department is responsible for the project and has co-ordinated meetings of the Project Team during the conceptual development of the project.

The outcome

After consideration of various options for the project, the installation of bulk silos on site was selected as the most appropriate option. It provided the best technological solution in terms of achieving operation exposure limits. It also provided an opportunity to reduce headcount providing a payback and financial justification for the project.

The project was executed on time, with handover completed during product changeover. Operators were trained by project department staff on completion of the project during the commissioning phase. Initial material deliveries arrived on time for commissioning and the overall supply chain continued with minimal interruption (Tables 3-9 and 3-10).

Table 3-9 A completed 'Why?' Checklist — case study C

Project Management Toolkit — 'Why?' Checklist

Project: Bulk powder handling	**Project Manager:** Confidential
Date: Confidential	**Page:** 1 of 2

Sponsorship

Who is the sponsor? (The person who is accountable for the delivery of the business benefits)
Production Manager
Has the sponsor developed an external communication plan? (How the sponsor will communicate with all stakeholders in the business)
The project timescale is approximately 6 months. The Production Manager will make the other site departments aware of the project at the next site management meeting. This meeting will be used to provide updates as the project progresses

Business benefits

Has a business case been developed?
The main project driver was a health and safety issue, as the material has now been identified as a sensitizer and present bulk bag arrangements for the material cannot meet exposure requirements. Occupational exposure limits have been set by the HSE
Have all benefits been identified? (Why is the project being done?)
The new facility must achieve required exposure limits for the powder material
The handling and unloading of bulk bags is a full time role within the manufacturing facility, a change to bulk handling with silos and automated transfer systems will allow a reduction in this headcount as the Operator controlling blending operation which is fed with the solid will be able to control both operations. The installation of a bulk facility will reduce the amount of raw material movements and deliveries on site, however the benefits from these changes have not been included within the business case
Who is the customer? (Identify all stakeholders in the business including the customer)
The production department is the main customer for the project. Other key stakeholders include, site engineering, stores, purchasing, site security as well as production management, engineers and operators
How will benefits be tracked? (Have they been adequately defined?)
On completion of project handover and training the resulting headcount reduction should allow production capacity to remain at present levels. This will be tracked through monitoring of overtime requirements. The required exposure requirements will be tested via environmental monitoring during commissioning of the equipment. Long-term health monitoring for all employees is a normal part of site operations (Table 3-11)

Business change

Will the project change the way people do business? (Will people need to work differently?)
Yes, purchasing will need to negotiate a bulk supply of material, warehouse will not be required to unload, store and deliver the material. Bulk tankers will have to be allowed into the site local to the manufacturing plant to make deliveries rather than to the main site warehouse. The change to headcount and training requirements will need to be handled appropriately so that the motivation of the manufacturing team is not impacted
Is the business ready for the project? (Are training needs identified or other organizational changes needed?)
The project has been discussed with purchasing and warehousing departments during its conceptual development. Bulk supplies of the powder material are available. Access to the production area is available. Training requirements will be extensive for plant operators as the bulk system will include a more advanced control system than anything present within the production area at present

(continued)

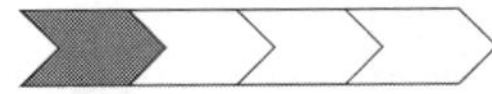

Table 3-9 (Continued)

Project Management Toolkit — 'Why?' Checklist	
Project: Bulk powder handling	**Project Manager:** Confidential
Date: Confidential	**Page:** 2 of 2
Scope definition	
Has the scope been defined? (What level of feasibility work has been done?) Various options have been considered for the project and presented to key stakeholders. The installation of bulk silos for material storage has been selected as the most appropriate technology for the site **Have the benefit enablers been defined? (Will the project enable the benefits to be delivered when the project is complete?)** The health and safety requirements are well defined as is the level of automation required in order to allow the reduction in headcount **Have all alternatives been investigated? (Which may include not needing the project)** The health and safety issue does not allow no action to be taken. Several options have been considered for the project including the use of personnel protective equipment to prevent operator exposure, modifications to the existing bulk bag arrangement or the use of solid bulk containers. The use of bulk silos has been selected as it is able to provide the desired exposure limits for the material and it reduces material movements allowing the desired headcount reduction **Have the project success criteria been defined and prioritized?** ➤ Plant design must contain powder and restrict release to atmosphere within designated limits ➤ Plant automation must allow a decrease in operator headcount	
Stage One decision	
Should the project be progressed further? (Is the business case robust enough for detailed planning to commence?) Yes — the idea appears to be linked to a real organizational need and the supplementary tools to support the benefits and business case definition validate this. See attached Simple Benefits Hierarchy (Figure 3-9) and Business Case Tool (Table 3-12)	

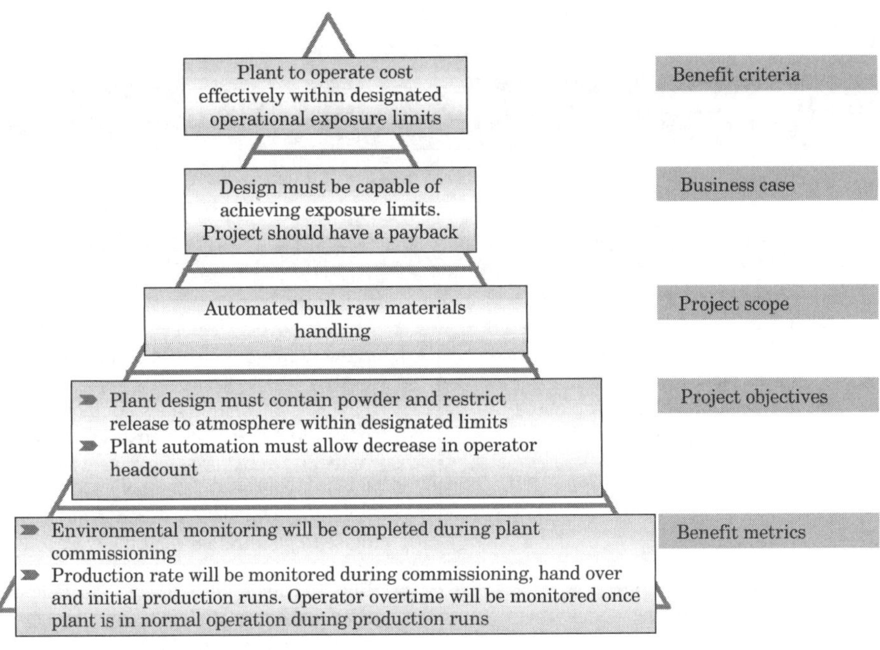

Figure 3-9 A completed Simple Benefits Hierarchy — case study C

The benefits map shown in Figure 3-10 supported the development of the Benefits Specification Table (Table 3-11).

Table 3-10 A completed Detailed Benefits Hierarchy — case study C

***Project Management Toolkit* — Detailed Benefits Hierarchy**		
Project: Bulk powder handling	**Date:** Confidential	
The project will install bulk silos for powder raw material storage adjacent to the manufacturing facility. Material deliveries will be made directly from supplier road tankers into these silos. Transfers into the manufacturing plant will be made by fully contained conveying systems. The entire system will be automated		
Level in hierarchy	**Item in hierarchy**	**Comment**
1 Benefit criteria	➤ Plant to operate within designated operational exposure limits ➤ Reduced operation costs for the facility by reduction in headcount ➤ Reduction in materials movements and warehousing requirements	➤ Achievement of operational exposure limits is the major driver for the project. This is a statutory and non conformance will prevent long-term operation of the facility ➤ Materials movements and reduction of warehousing requirements are difficult to measure and will not be included within project justification

(continued)

Table 3-10 (Continued)

Project Management Toolkit — Detailed Benefits Hierarchy		
Project: Bulk powder handling		**Date:** Confidential
Level in hierarchy	**Item in hierarchy**	**Comment**
2 Business case	➤ Design must provide designated exposure limits during operation of the plant ➤ Automation of the new plant should allow a reduction in headcount	➤ Project payback calculation is based on head count reduction and no increase to production overtime costs ➤ Payback calculation does not include realization of warehouse space or costs associated with the reduction in material movement around the site ➤ Inventory levels for the raw material will not change. Inventory will just be stored in another place (bulk silo close to plant rather than the warehouse)
3 Project scope	➤ Automated bulk raw materials handling	➤ Bulk raw materials storage — install bulk storage silos and automated filling and transfer equipment to existing blending plant
4 Project objectives	➤ Plant design must contain powder and restrict release to atmosphere within designated limits	➤ Although the main criterion for design success is compliance additionally the design has to consider operation of the supply chain
	➤ Plant automation must allow decrease in operator headcount	➤ As the original bulk bag operation was an extremely labour intensive process, the option of providing automation should allow a reduction in headcount for the plant operating team ➤ The automation is a CSF — without it the reduction in headcount, and therefore manufacturing cost, is not possible. Additionally production capacity must not be affected
5 Benefit metrics	➤ Environmental monitoring during project commissioning	➤ This monitoring is infrequently carried out, usually on an annual basis. Previous data is available
	➤ Production rate will be monitored during commissioning and handover	➤ This is a normal operating measure, daily and monthly trends are available
	➤ Headcount reduction	➤ This must be measured to ensure that the unnecessary headcount is reallocated to another role where he is needed
	➤ Production overtime will be monitored following handover	➤ This is normal operating measure, monthly trends are available. This is to be monitored to check that the reduced headcount has not meant additional overtime and is linked to the success of the automation part of the project

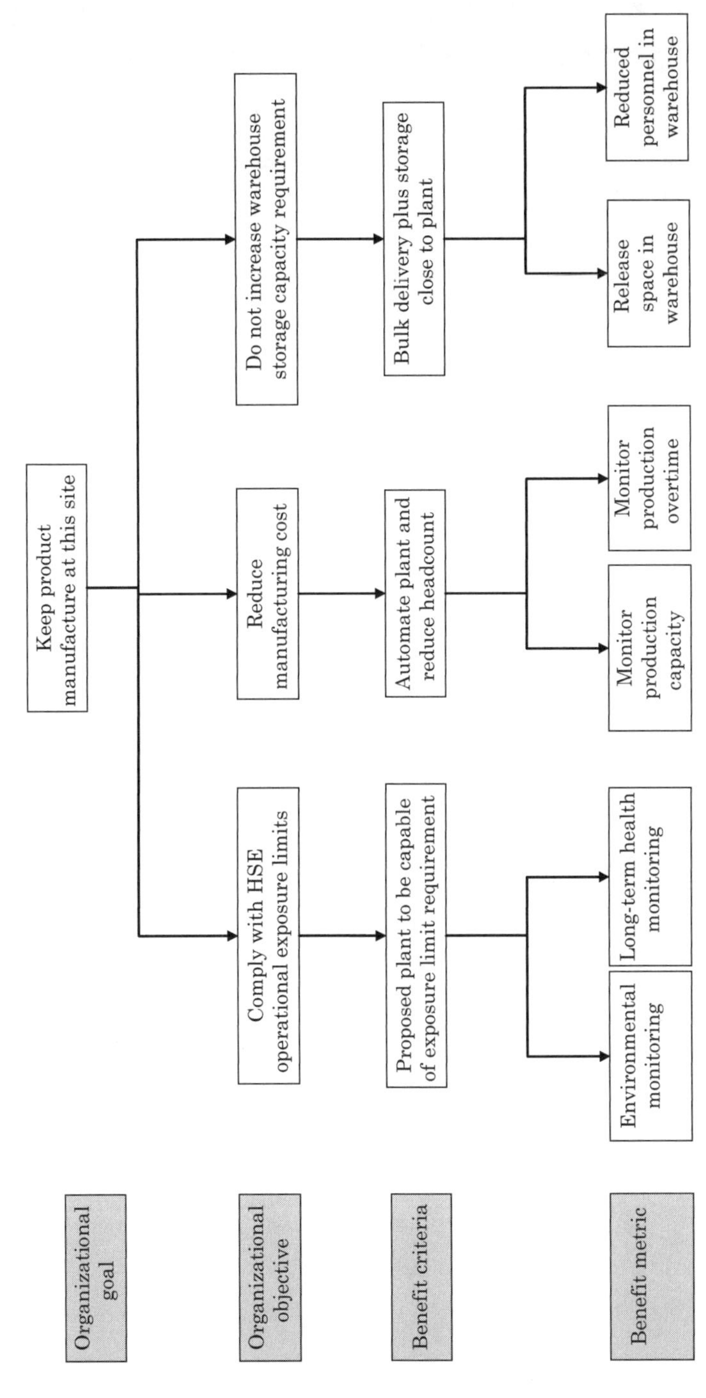

Figure 3-10 A benefits map — case study C

Table 3-11 A completed Benefits Specification Table — case study C

Project Management Toolkit — Benefits Specification Table

Project: Bulk powder handling			**Date:** Confidential		
Potential benefit	**Benefit metric**	**Benefit metric baseline**	**Accountability**	**Benefit metric target**	**Sub-project**
(What the project will enable the business to deliver)	*(Characteristic to be measured)*	*(Current level of performance)*	*(The person who is accountable for delivery of the benefit to target)*	*(Required performance to achieve overall benefits)*	*(The project scope that will enable this benefit to be delivered)*
Limit operator exposure within designated limits	Environmental monitoring	Current occupational exposure levels	Process Design Engineers	HSE designated limit	Plant equipment specification and equipment selection
Lower manufacturing cost through automation for same production throughput	Headcount reduction	Current headcount	Production Manager	Headcount to reduce by one	Production Manager to relocate surplus staff member after plant automation project completion
		Current production capacity	Production Manager	Production capacity to remain at present level	Plant automation
Bulk delivery and storage (no longer use bulk bags)	Reduced warehousing requirements (space)	Space currently occupied by raw material	Materials Manager	Existing holding of bulk bags should be reduced to zero	Installation of bulk storage and offloading equipment
	Reduced workload for warehouse personnel	Warehouse personnel time to store and transfer bulk bags (accurate data not available)	Materials Manager	A claimed benefit will not be made for project justification Improvements will be monitored following project completion	Installation of bulk storage and offloading equipment

Table 3-12 A completed Business Case Tool — case study C

***Project Management Toolkit* — Business Case Tool**			
Project: Bulk powder handling		**Date:** Confidential	
Business case developed by	Confidential	**Date**	Confidential
Project reference number	Confidential	**Business area**	Confidential
Project Manager	Confidential	**Project sponsor**	Confidential
Business background	Main production unit blending facility upgrade of powder raw material handling from bulk bags to bulk silo storage. The upgrade is required in order to achieve new occupational exposure limits which are a statutory requirement		
Project description	The project will install bulk silos for powder raw material storage adjacent to the manufacturing facility. Material deliveries will be made directly from supplier road tankers into these silos. Transfers into the manufacturing plant will be made by fully contained conveying systems. The entire system will be automated		
Delivery analysis	The site project department engineering team will complete the project, with support from external design contractors where suitably qualified internal staff are not available. Key stakeholders in the project are the purchasing, materials management and manufacturing departments. Purchasing have already confirmed that bulk tanker deliveries are available and will need to ensure that these arrive in the correct form with the correct fittings on completion of the project in order to allow commissioning to take place. Materials management (warehouse team) will no longer be required to store and transfer bulk bags and should ensure that bulk bag storage is reduced as far as possible prior to the change over to the new system. Manufacturing will operate the new equipment when it is complete		
Business change analysis	➤ Statutory occupational exposure limits must be achieved within the plant in order to allow long-term future operation ➤ The automated bulk storage facility will allow a reduction in headcount within manufacturing ➤ The warehouse space currently occupied by bulk bag storage will be available for something else ➤ Materials management and goods receipt will no longer be required to offload, and move bulk bags around the site (warehouse team)		
Value-add analysis	➤ Occupational exposure limits must be achieved for future plant operation ➤ Headcount reduction will provide reduced manufacturing costs ➤ Changes to materials management activities and warehousing are an obvious benefit but is difficult to quantify. An assessment should be made by the warehouse team following completion of the project		
Impact of not doing the project	There is potential for long-term health risks to plant operational staff and frequent visitors to the plant such as engineers, technicians and analytical staff. Increasing levels of personal protective equipment could be utilized to protect employees, however the facility was never intended to operate this way and is not equipped with changing and clean down areas		
Project approved *(Value-add or not?)*	Yes/~~No~~ *<note that this approval approves all the above requests>*	**Name of approver and date**	Confidential

Handy hints

Never be afraid to ask 'why?'

By the time most Project Managers get involved with a project the initial justification work has been completed and the project is simply handed over for delivery. Even at this stage it is critical that the Project Manager understands why the project is required.

Don't progress your project without a 'real' sponsor

A project needs to have a sponsor — someone who is accountable for the realization of the business benefits. Many projects have failed to deliver what was expected of them because of this lack of organizational ownership.

Do link your project to the goals of your business

No businesses have the luxury of authorizing projects which have no apparent use other than to keep a Project Manager and his team busy whilst spending the company's money — all 'real' projects have an organizational purpose. If you can't link the project to the goals of your business then it isn't a project you should be doing.

Most projects will change some part of the organization — find out where and how

Business change project management is often seen as a separate project management discipline — business change projects are seen as being different from engineering or capital based projects. However most projects (any type) will change the business in some way and so most projects require some business change to be managed as well. It's important that you know where and how your project will change the organization and also to ensure that someone (if it's not you) is sorting out that part of the project.

Always look to expand your Stage One toolkit

The Stage One toolkit contains four tools out of many that are actually used in the early stages of project definition when the project is no more than an idea. There are other tools which will help you as you define and refine your understanding of what the project needs to be to deliver a specific set of benefits for your business — search them out and use them appropriately and pragmatically.

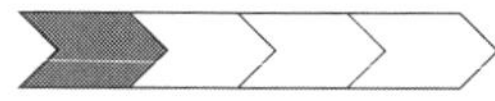

Further reading

Benefits management and business change management are clearly crucial parts of project management, as is taking a more holistic approach to project management. These aspects are covered in more detail by *Project Benefits Management: Linking Projects to the Business* a further book in this Project Management Series (ISBN: 978-0-75068477-4).

Additionally the websites referenced on page xi can be used to source further 'best practice' material.

And finally . . .

- Ask 'why?' — if you don't know why you're doing the project maybe nobody does!
- Understand how your project connects to the business — remember no project is an island!
- Be flexible — use and adapt the tools to achieve your goals.

4 Stage Two: how?

The second value-add stage in a project involves asking 'how?'

- How can the project be delivered?
- How can the organization be assured of being able to realize the benefits?

Every project requires a robust plan if it is to be successfully delivered and the benefits realized sustainably.

The misconception in project planning is that this relates to merely the time-based plan of activities to be completed. Whilst this is an important aspect of project delivery planning it is only one part.

How?

In the simplest terms — planning is the link between 'what we said we'd do' to 'what we actually deliver'. Figure 4-1 summarizes some of the main reasons why plans are needed. Planning links

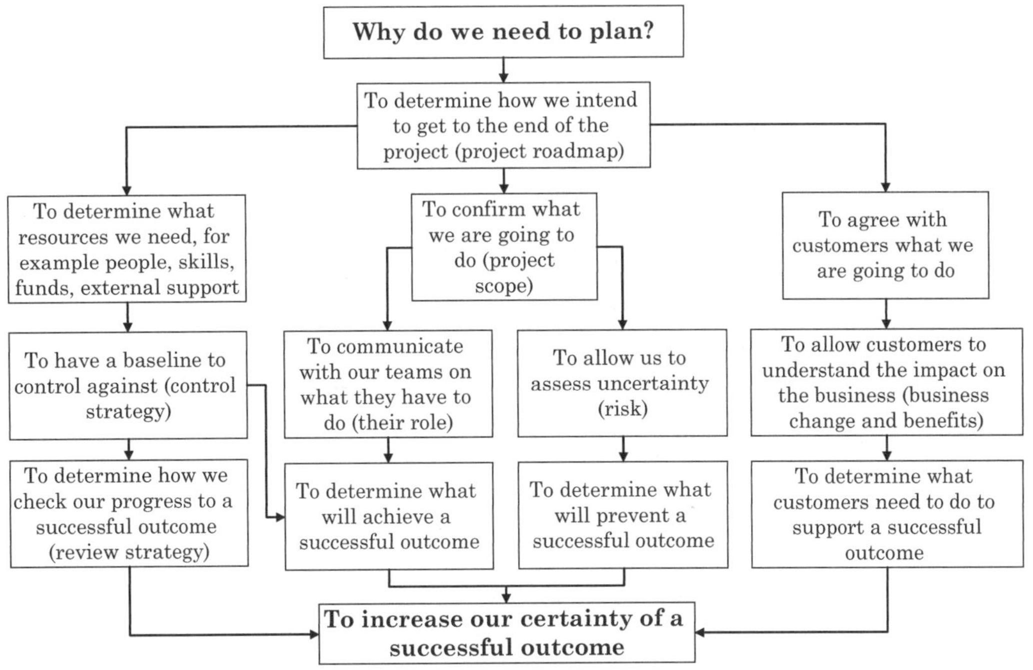

Figure 4-1 Why we plan

effective management of cost, time and scope with team management, development of a shared commitment to the outcome and continued reassurance that customers will get what they need.

Ultimately though planning is all about increasing chances of success within an uncertain world. Good project management is about effective management of uncertainty and project delivery planning is the mechanism for increasing the certainty of a successful outcome.

Project delivery planning

A project delivery plan is specifically prepared for a project based on standard project delivery planning practices. Its aim is to ensure that strategies specific to the needs of that project are developed, implemented, communicated and constructed around 'best practice' project management (Figure 4-2). It:

- Defines the project purpose and objectives, the Project Team organization with individual responsibilities and the project delivery strategies required to ensure that the project outcome is successful and meets the business needs.
- Is a key communication tool for the organization — all members of the Project Team, the sponsor and all other stakeholders within the business.

The project delivery plan is a 'live' document, which should be updated at regular intervals. This is necessary to ensure that as strategies are developed and refined, they are aligned to support achievement to the project objectives and business benefits. The initial project delivery plan should be developed and approved prior to the start of the project delivery phase. In a large or complex project a number of project delivery plans may be developed and linked through a hierarchy of project delivery plans (Figure 4-12 in Case Study F).

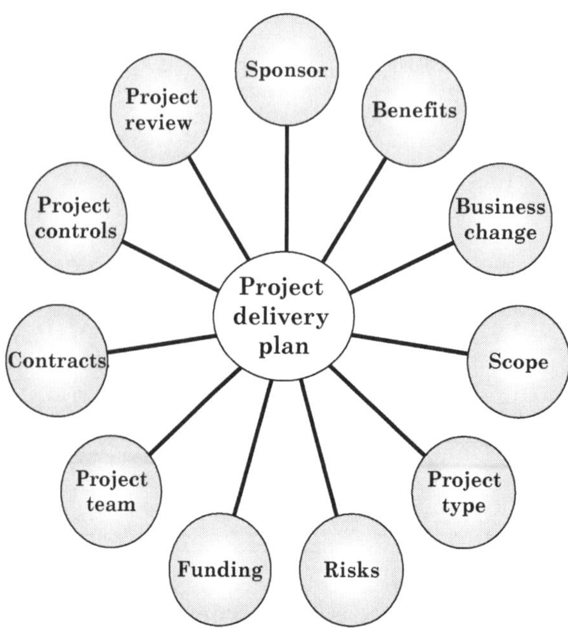

Figure 4-2 Components of a project delivery plan

In summary project delivery planning is a method to maximize success by planning to manage all areas of uncertainty.

The following tools are just a few which can be used during the planning stages of a project to support you in asking 'how?':

- 'How?' Checklist.
- Stakeholder Management Plan.
- Table of Critical Success Factors (CSFs).
- RACI Chart.
- Control Specification Table.

Projects which do not receive this level of planning at an early stage in their life are more likely to fail as:

- The project is more likely to become 'out of control' during delivery — if nobody knows where they're going it's quite hard to get there.
- Uncertainty won't be effectively managed — if there aren't plans to manage risks then when they appear they'll impact success.

Tool: 'How?' Checklist

If any project (at whatever stage) cannot be challenged by the questions contained in this checklist *and* have robust answers, suitably backed up, then it isn't robust enough to be delivered, that is to say it doesn't have a robust enough project delivery plan.

This tool can also be used proactively throughout Stage Two as a reminder of what should be done next.

The 'How?' Checklist explained

The checklist has thirteen major checks:

- Stage One check.
- Sponsorship.
- Benefits management.
- Business change management.
- Scope definition.
- Project type.
- Risk and issue management.
- Funding strategy and finance management.
- Project organization.
- Contract and supplier management.
- Project controls strategy.
- Project review strategy.
- Stage Two decision.

Table 4-1 shows the 'How?' Checklist with high-level guidance on how to complete each check. The rest of this section provides additional, more detailed, notes to support checklist completion.

Stage One check

This check is included as there is expected to have been significant work on the project since the end of Stage One. It is critical for the robustness of any project delivery plan that the link between the project and the business is checked so the following question needs to be asked:

- Have there been any changes since Stage One completion?

Additionally, the work which may have been completed on the scope development may have altered the business case or benefits realization plan; this should be reviewed.

Sponsorship

These two checks confirm that the sponsor remains the same as the person identified during Stage One and their role in effective communication:

- Who is the sponsor?
- Has the sponsor developed a communication plan?

Occasionally the sponsor may change during the early phases of a project as the benefits become clearer. Only the person who can take accountability for the delivery of the business benefits can be an effective sponsor.

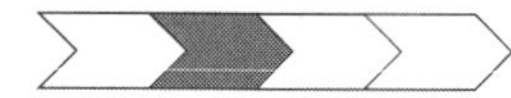

Table 4-1 The 'How?' Checklist explained

Project Management Toolkit — 'How?' Checklist

Project: *<insert project title>*	**Project Manager:** *<insert name>*
Date: *<insert date>*	**Page:** 1 of 3

Stage One check

Have there been any changes since Stage One completion? (Development of the business case and project kick-off may be some time apart)
<insert any changes that may impact the delivery of the project or the associated benefits>

Sponsorship

Who is the sponsor? (The person who is accountable for the delivery of the business benefits)
<confirm the name of the person who is taking this role>
Has the sponsor developed a communication plan?
<insert a comment on how the sponsor intends to communicate with project stakeholders during the project>

Benefits management

Has a benefits realization plan been developed?
<insert any data related to the schedule for delivery of the agreed benefit metrics>
How will benefits be tracked? (Have they been adequately defined?)
<insert any additional data which further articulates the specific benefit metrics which align with work completed during Stage One>

Business change management

How will the business change issues be managed during the implementation of the project? (Are there any specific resources or organizational issues?)
<insert specific plans for the management of the business change associated with the project>
Have all project stakeholders been identified? (Review the stakeholder map from Stage One)
<attach the stakeholder analysis work that has been completed>
What is the strategy for handover of the project to the business? (Link this to the project objectives)
<insert specific plans for project handover>

Scope definition

Has the scope changed since Stage One completion? (Has further conceptual design been completed which may have altered the scope?)
<insert details of the further work which may have been conducted prior to project kick-off>
Have the project objectives been defined and prioritized? (What is the project delivering?)
<Attach a copy of the prioritized objectives>
<insert an updated list of project CSFs>

Project type

What type of project is to be delivered? (For example engineering or business change)
<insert the type of project being delivered — note that this is a major category>
What project stages/stage gates will be used? (Key milestones for example funding approval, which might be go/no go points for the project)
<insert the project roadmap for the type of project within the organization>

(continued)

Table 4-1 (Continued)

Project Management Toolkit — 'How?' Checklist	
Project: *<insert project title>*	**Project Manager:** *<insert name>*
Date: *<insert date>*	**Page:** 2 of 3
Funding strategy and finance management	
Has a funding strategy been defined? (How will the project be funded and when do funds need to be requested?) *<insert the funding request requirements — estimate accuracy, funding timeline, authorization process>* **How will finance be managed?** *<confirm that no additional reporting outside of the project control strategy is required>*	
Risk and issue management	
Have the CSFs changed since Stage One completion? (As linked to the prioritized project objectives and the critical path through the project risks) *<insert updated critical path of success if available>* **Have all project risks been defined and analysed? (What will stop the achievement of success?)** *<comment on any high priority risks>* **What mitigation plans are being put into place?** *<attach a copy of the high priority mitigation plans>* **What contingency plans are being reviewed?** *<attach a copy of the high priority contingency plans>* *<attach a copy of the Risk Table and Matrix>*	
Project organization	
Who is the Project Manager? *<insert the name of the Project Manager who will be delivering the project in line with the project delivery plan>* **Has a project organization for all resources been defined? (Include the Project Team and all key stakeholders)** *<insert any comments on the project resource situation — capacity or capability>* *<have roles and responsibilities been defined? Attach the RACI Chart and/or project organization chart>*	
Contract and supplier management	
Has a strategy for use of external suppliers been defined? (The reasons why an external supplier would need to be used for any part of the scope) *<insert a copy of the contract plan>* **Is there a process for using an external supplier? (For example selection criteria, contractual arrangements, performance management)** *<confirm that procedures to manage supplier selection and performance are in place>*	
Project controls strategy	
Is the control strategy defined? *<comment on each of the following:* ➤ *Cost control strategy* ➤ *Schedule strategy* ➤ *Change control* ➤ *Action/progress management* ➤ *Reporting* *What methodologies, tools or processes will be used to ensure control?>*	

(continued)

Table 4-1 (Continued)

<table>
<tr><th colspan="2">Project Management Toolkit — 'How?' Checklist</th></tr>
<tr><td>Project: <insert project title></td><td>Project Manager: <insert name></td></tr>
<tr><td>Date: <insert date></td><td>Page: 3 of 3</td></tr>
<tr><td colspan="2">Project review strategy</td></tr>
<tr><td colspan="2">Is the review strategy defined? (How will performance be managed and monitored — both formal and informal reviews and those within and independent to the team?)
<comment on the plan for reviewing project performance during the delivery of the project></td></tr>
<tr><td colspan="2">Stage Two decision</td></tr>
<tr><td colspan="2">Should the project be progressed further? (Is the project delivery strategy robust enough for project delivery to commence?)
<insert the decision — yes or no — with comments on the robustness of the project delivery plan></td></tr>
</table>

During the planning phases of a project the Project Manager and the sponsor will need to work together in order to agree how best to communicate with all project stakeholders. Stakeholder management planning can support this activity.

Benefits management

These two checks confirm the current status of the benefits delivery planning:

- Has a benefits realization plan been developed?
- How will benefits be tracked?

A benefits realization plan includes a schedule for the delivery of the agreed benefit metrics based on a clear definition of each metric. At this stage in the project planning it is important to work with the sponsor to develop a realistic plan which is related to the project time-line:

- When an organization expects delivery of benefits (in timing and quantity) will impact on the project scope (in timing and quantity).

Business change management

On the basis that the link to the business has already been established during the earlier stages of the project and the required business changes are known, these checks establish whether the delivery of the changes in the business have been planned:

- How will the business change issues be managed during the delivery of the project?
- Have all project stakeholders been identified?
- What is the strategy for handover of the project to the business?

The customer for a project will usually have some operational changes to deliver in order to be ready for the completion and handover of the project, for example training, re-organization, recruitment, and procedural changes. These will require resources and need to be planned. As a part of the Stakeholder Management Plan the Project Team may choose to support or influence this planning activity.

Developing a plan to manage these external project issues is important and is based around the Project Team's understanding of all project stakeholders. Stakeholder planning and the Stakeholder Management Plan are discussed further on page 64.

Scope definition

As already mentioned — the scope may have been further detailed through specific feasibility work since approval of the business case. Therefore these checks are aimed at confirming that the approved scope is still being delivered as this links to the approved business case and benefits realization plan:

- Has the scope changed since Stage One completion?
- Have the project objectives been defined and prioritized?

At this stage it is appropriate to attach a copy of the prioritized objectives. These should accurately reflect what the project will be delivering in terms of quantity, quality and functionality. Additional delivery criteria related to timing may also be important.

Scope must be well defined and agreed in terms of quantity, quality and functionality as it is the basis for:

- The development of a sound control strategy.
- Funding requests.
- An understanding of the technical skills required within the project organization.
- A robust risk assessment.

One methodology which can support the effective development of the detailed scope is the development of a work breakdown structure — packages of scope and sub-scope. An example of a tool that can support this is the Table of Critical Success Factors, which is described on page 70.

Project type

These checks address whether the project roadmap has been developed — the route that the project must travel in order to achieve success:

- What type of project is to be delivered? (For example engineering or business change)
- What project stages/stage gates will be used? (Key milestones for example funding approval, which might be go/no go points for the project)

Each type of project has a roadmap that describes the optimum stages and stage gate reviews that the project should go through to achieve its objectives and therefore success. A project roadmap defines the processes that are to be used to progress through the various stage gates such as:

- Information systems projects — design, procurement and configuration management.
- Product development projects — disease and target selection, candidate and preclinical development phase I and II trials.
- Business change projects — data collection, problem diagnosis, design, delivery, sustainability.
- Engineering projects — design, procurement, construction and commissioning (Figure 4-3).

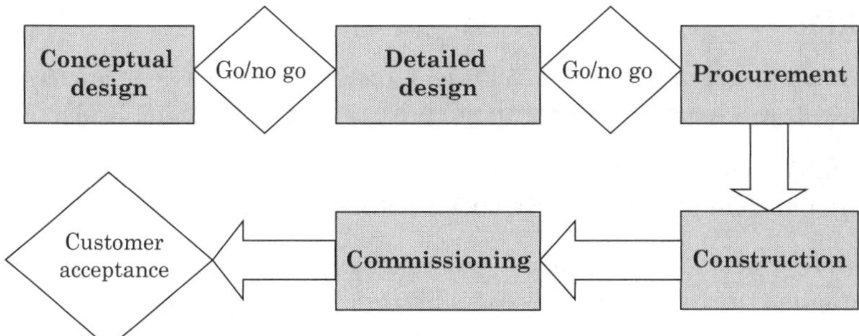

Figure 4-3 An example project roadmap — engineering project

Funding strategy and finance management

These checks address whether the way that the project will be funded has been worked out, for example who will be asked for funds (capital or revenue) and when does this need to be done:

- Has a funding strategy been defined?
- How will finance be managed?

In most organizations the process to request capital funding is complex and relies on accurate funding estimates against a robust scope and business case. The appropriate process within an organization needs to be identified and then incorporated into the project delivery plan. The management of any approved finance (capital or revenue) is usually incorporated into the project control strategy — cost, schedule and scope control — as it concerns the delivery of an investment against agreed funds within an agreed timescale.

Risk and issue management

All projects must define and manage risk. This is a key part in the process of identifying and managing uncertainty. At this stage a plan to manage risk must be developed and the following four checks challenge this plan:

- Have the CSFs changed since Stage One completion?
- Have all project risks been defined and analysed? (What will stop the achievement of success?)
- What mitigation plans are being put into place?
- What contingency plans are being reviewed?

It is important to understand what is critical to success as it is the first step in understanding what could prevent success. A standard method to assess and manage risks is by use of the Risk Table and Matrix (see page 119). This tool supports identification of mitigation plans (eliminate risk or reduce impact) and contingency plans (back-up plan when the risk occurs).

When a risk occurs it becomes an issue and an 'action plan' methodology is usually employed to manage the issue, the contingency plan, until resolution.

Project organization

A Project Manager must be appointed and that Project Manager must have a plan to select and manage all individuals on the project in order to have an effective, highly performing Project Team:

- Who is the Project Manager?
- Has a project organization for all resources been defined?

It is all too easy to rely on standard Project Team organograms when doing a certain type of project. However the most appropriate methodology is to build team structure around the required team capacity and capability linked to scope, cost and schedule requirements.

A proposed method to achieve this is based around the use of the RACI Chart (see page 75).

Contract and supplier management

As a part of the development of the project organization it will become clear if external support is needed. If so, the Project Manager must have a plan to select and manage external suppliers:

- Has a strategy for use of external suppliers been defined?
- What is the process for using an external supplier? (For example selection criteria, contractual arrangements, performance management)

Most organizations will have a robust purchasing function however the Project Manager must be able to clearly articulate why that external supplier is needed. Any contract management support from other parts of the organization (for example procurement and legal) must always be supplemented with supplier performance management from within the project:

- A contract supply plan should be agreed before kick-off (this is like a mini-delivery plan just for the supply of the external scope).
- Appropriate performance management should be used (this is like a mini-control strategy for the external scope addressing the required cost, schedule and scope management).
- A formal contract close-out is needed (this is a final contract review).

Any external providers are a part of the overall project organization and should be considered within a high-level Project Team organogram and associated RACI Chart.

Project controls strategy

A major part of any project delivery plan is working out the best way to manage the uncertainty so that the certainty of outcome is increased. An inappropriate control strategy can be as detrimental to a project as not having a control strategy. This check is therefore reviewing what tools, techniques and processes are going to be used on the project:

- Is the control strategy defined?

The control strategy usually addresses the management of amongst other things cost, schedule, scope, change, delivery progress and reporting.

The Control Specification Table (see page 79) and the Earned Value Tool (see page 124) are both useful in addressing this key issue.

Project Review Strategy

This check is challenging whether the Project Manager has determined how project performance will be managed and monitored:

- Is the review strategy defined?

Within the plan it is usual to highlight the frequency of:

- Project Manager meetings with the sponsor.
- Project Team meetings.
- External project reviews (which may be a part of the organizations overall project governance procedures).

Additionally, any formal stage gate review as required by the project roadmap will constitute a formal project review.

Stage Two decision

Once all the available project delivery planning information has been gathered it is critical that the Project Manager, in partnership with the project sponsor, makes a clear decision on the robustness of the work to date and asks:

- Should the project be progressed further?

The end point for Stage Two is an agreement that the project delivery strategy is robust enough for the project delivery to commence. This does not mean that all planning activities are complete because within Stage Three, Project Delivery, the monitoring and reviewing of the plan is a key activity.

Using the 'How'? Checklist

This is a key tool for the Project Manager that can be used in a number of ways:

- To prompt the Project Manager as the project delivery plan is being developed.
- To check the robustness of the plan at the end of the planning phase prior to commencement of delivery.
- To gain Project Team buy-in and shared commitment to the project goal.
- To support effective communication with the sponsor and other senior stakeholders — the project delivery plan can act as a 'contract' between stakeholders.

In the latter mode it is best completed in partnership with the particular stakeholder, although with the previous two modes it will more than likely be used by the Project Manager. The ultimate aim of the checklist, however, is to answer one question:

- Does the project have a robust plan? (Stage Two decision)

To complete the checklist the Project Manager will need access to the completed project delivery plan which will contain:

- 'Hard' data such as statements of scope, risk and team responsibilities.
- 'Soft' data such as processes for internal and external stakeholder management and communication.

Each of the case studies in this chapter uses the checklist to demonstrate the different ways the tool can be used.

In completing the 'How?' Checklist it can be seen how the use of other tools within this chapter can support the development of a robust project delivery plan and therefore the completion of the Stage Two decision.

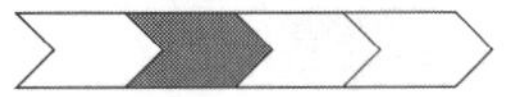

Tool: Stakeholder Management Plan

The aim of this tool is to support the effective delivery of a project by identifying and understanding the potential impact of all project stakeholders:

- Internal project stakeholders — Project Team members.
- External project stakeholders — sponsors, customers and end users.

Traditionally organizations have focused on few links between the project and the organization in order to maintain control and focus. However today it is recognized that a project can be impacted by the behaviours, perceptions and actions of a wider group of people and that this needs to be managed.

The Stakeholder Management Plan explained

The tool contains two main sections:

- Individual stakeholder analysis.
- Summary stakeholder analysis.

The Stakeholder Management Plan is shown in Table 4-2 with high-level guidance on how to complete each section. The following are additional notes to support tool completion.

Table 4-2 The Stakeholder Management Plan explained

***Project Management Toolkit* — Stakeholder Management Plan**						
Project: *<insert project title>*				**Date:** *<insert date>*		
Individual stakeholder analysis						
Stakeholder name/role	**Type**	**Current project knowledge**	**Level of influence**	**Current level of engagement**	**Target level of engagement**	**Management of stakeholder**
<insert stakeholder's name and role in the organization>	*<insert type>*	*<insert low, medium or high>*	*<insert low, medium or high>*	*<insert low, medium or high and a comment on the evidence to back this up>*	*<insert low, medium or high and a comment on the actions needed>*	*<insert project team member's name>*
Summary stakeholder analysis						
<insert comments on the overall impact of the current stakeholder situation on the project>						

Individual stakeholder analysis

Stakeholder name/role

Initially all stakeholder names are listed in the first column in the section. It is important that their role in the organization is noted because for some stakeholders this will be the reason they are in the plan. For example, if the person changes role during the life of the project then it is the new person who becomes the stakeholder, even if the original person maintains some influence, he has transferred power. The remaining columns then need to be completed for each stakeholder.

Type

This defines the person's role in the project, whether they are sponsoring the project, involved in the project (customer representative) or just impacted by the project (customer end user). Use the categories in Table 4-3 to complete this section.

Current project knowledge

All key stakeholders need to have some level of knowledge about the project. This enables assessment of what each person is likely to know.

Table 4-3 Stakeholder types

Stakeholder type	Description	Comment
Champions C	Want the project and attempt to obtain commitment and resources for it	Would not typically have the organizational power to support the project although they may be able to influence. Could also be a change agent or target
Change agents A	Will be needed to plan and implement the business changes which arise as a result of the project	Would typically be within the customer organization. Sometimes these are seconded onto Project Teams to support the effect link back to the business
Authorizing sponsor S(a)	Ownership of the project	Would be the most senior person from an organizational perspective — he controls the targets impacted by the project
Reinforcing sponsor S(r)	Local ownership of the project	Would be a senior person within one part of the organization impacted by the project
Targets T	Those impacted by the project. They may have to work differently as a result of the project, for example learn new skills and follow new procedures	Would be anybody within the organization impacted by the project. These are the end users who may also be brought into the Project Team

Level of influence

An assessment of whether this person can influence the outcome of the project, either because of their organizational authority or their informal position within the business area.

Current and target level of engagement

Engagement here refers to the level of real support that a stakeholder has for a project. For example, you would want a customer to be highly engaged in a project he requested but if he has other demands on his time you may not get it. Additionally if they don't know enough about the project then they are also unlikely to be engaged — communication is a key stakeholder planning action for a Project Team.

Management of stakeholder

Each person should be assigned to a member of the Project Team, for example the Project Manager will manage the sponsor and may also be allocated a senior customer representative who is now a part of the Project Team. The role here involves ensuring that all required stakeholder actions are completed.

Summary stakeholder analysis

The final section in the Stakeholder Management Plan allows for an analysis of the overall stakeholder situation:

- Have any serious concerns been identified which are potential risks to the project success?
- Are there any 'barriers' within the stakeholder group who are already having an impact on project success?

Using the Stakeholder Management Plan

The basis for the generation of the Stakeholder Management Plan is a stakeholder mapping session. This is a method which identifies and categorizes all potential project stakeholders. The outcome of this working session is directly input to the Stakeholder Management Plan.

Stakeholder mapping is an activity which considers the influence specific people in an organization, Project Team or role, may have on the outcome of a project.

A stakeholder mapping session is best conducted with a small sub-set of the Project Team and can be facilitated by the Project Manager. Figure 4-4 shows a generic stakeholder map — an organization chart with key stakeholders highlighted.

In order to generate a stakeholder map the sub-set of the Project Team should be brought together in a 'facilitated brainstorm' where:

- The group should be asked 'which teams or organizations are involved in the project or impacted by the project?'
  - For each identified organization the group should be asked to develop an organization chart on a flip chart noting names and roles of key people.
  - The formal organizational links need to be correct but any particularly strong informal or formal links between people or roles or teams should be noted (dotted lines should be used for informal).

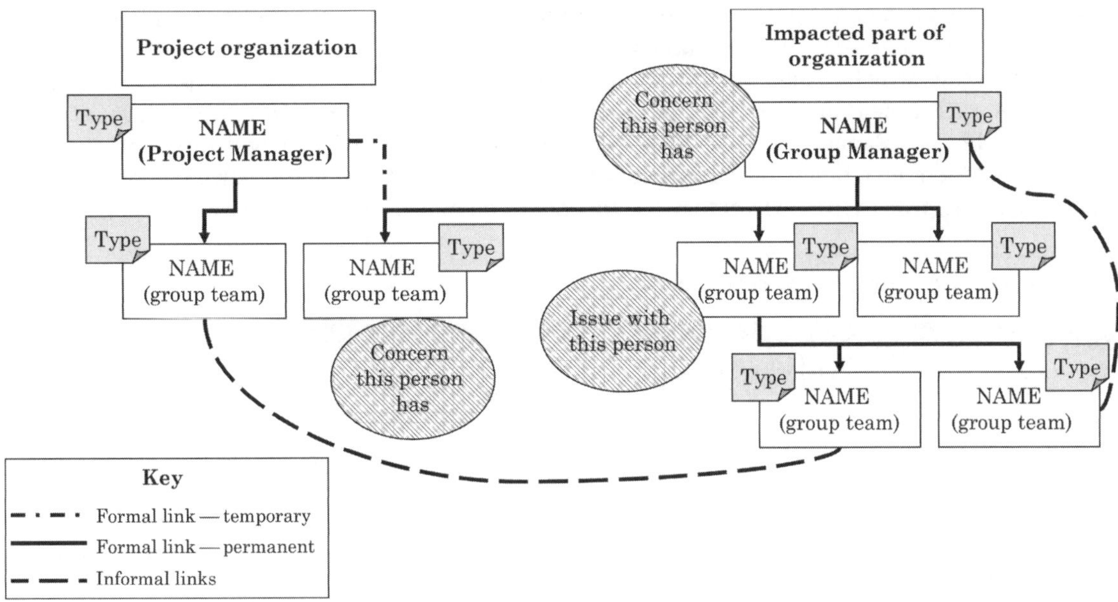

Figure 4-4 A generic stakeholder map

- The group should be asked 'who supports this project and who doesn't?' and they should write their responses on Post-it® notes:
  - The Post-it® notes should be placed on the flip chart next to the person's name.
  - The group may agree or disagree on the stakeholder position — any disagreements should be noted for discussion later.
  - The group should add specific Post-it® notes if they have evidence of a specific behaviour (supporting or non-supporting).
- The group should be asked 'who are most important and why?' and they should write their responses on Post-it® notes:
  - The Post-it® notes should be placed on the flip chart next to the person's name.
  - The group may agree or disagree on the level of power or influence a particular stakeholder may have.
  - The group should add specific Post-it® notes if they have evidence of a specific behaviour (influence, power or the converse).
- The group should be asked to categorize the stakeholders according to the types listed in Table 4-3.
  - The group may agree or disagree on the stakeholder type.
  - The group should add specific Post-it® notes backing up their choices.
- The completed map should be reviewed and amended if necessary.
  - All disagreements need to be sorted out — this may involve collecting more evidence.
  - The stakeholders that are to be transferred to the Stakeholder Management Plan need to be agreed.

The Stakeholder Management Plan can then be populated and used in further discussions with the sub-set of the Project Team to confirm exactly how stakeholders are to be managed. As the project progresses the plan can be used as a 'live' document to note changes in stakeholders, and/or stakeholder behaviour, and therefore assess the potential impact on the project.

Any major issues which result from the stakeholder analysis will need to be discussed with the project sponsor as some issues may only be resolved by someone with the appropriate level of authority — a Project Manager will usually only have influence over external stakeholders.

Table 4-4 is an example of a part of a Stakeholder Management Plan generated from an extremely complex map. The project from which this was taken was an improvement project within a

Table 4-4 An example Stakeholder Management Plan

***Project Management Toolkit* — Stakeholder Management Plan**						
Project: *Cycle time reduction project*				**Date:** *Confidential*		
Individual stakeholder analysis						
Stakeholder name/role	**Type**	**Current project knowledge**	**Level of influence**	**Current level of engagement**	**Target level of engagement**	**Management of stakeholder**
Fred (Site Director)	S(a)	Low	High	Low — he keeps cancelling his meetings with the Project Manager	High — need him to release resources and to sort out some of the political issues between departments	Trish (Project Manager)
Bill (Production Manager)	S(r)	High	Medium	Medium — he is getting annoyed with the other department managers who keep finding blockers to the project	High — needs to use his enthusiasm to convince the other department managers and his own team that this project will deliver	Trish (Project Manager)
Sally (Planning Manager)	S(r)	Low	High	Low — she cannot see how the project will work and is completely disengaged	High — supply chain planning is a crucial process and the project will significantly change how this department works	Bob (Senior Design Engineer)
John (Planner)	C, A	High	Low	High — he has been going on about the need for this project for some time	High — his good relationship with Sally needs to be put to good use	Bob (Senior Design Engineer)
Summary stakeholder analysis						
Currently there are a number of issues that could potentially slow down the delivery of the project. If the design work is to be completed on time significantly more data collection is needed within each department and this would be best done with active help from the people in those departments						

manufacturing site looking specifically at improving the supply chain cycle time. The project impacted all departments in the supply chain of a key product and involved engineering, procedural and cultural changes. The stakeholder map was used to understand the key characters within the organization and the plan developed effectively managed the risks that this had highlighted.

Stakeholder management plans should be treated as confidential documents due to the nature of the data they contain.

Tool: Table of Critical Success Factors (CSFs)

The aim of this tool is to challenge the project's scope to ensure that any activity or deliverable required for project success is included. Any scope, which does not contribute to the project outcome, is eliminated.

It is based on a methodology of identifying a critical path within the scope and then detailing a work breakdown structure as a series of more and more detailed critical success factors (CSFs).

The Table of Critical Success Factors explained

The tool contains two main sections:

- Critical path of success.
- Critical success factor definition.

Table 4-5 shows the Table of Critical Success Factors with high-level guidance on how to complete each section. The following are additional notes to support tool completion.

Table 4-5 The Table of Critical Success Factors explained

***Project Management Toolkit* — Table of Critical Success Factors**				
Project: *<insert project title>*		**Date:** *<insert date>*		
Critical path of success				
<insert a brief description of the intended project outcome which requires the completion of the Level 1 critical success factors (CSF1)>				
Critical success factor definition				
Scope area (CSF Level 1)	**Objective tracking metric (CSF Level 2)**	**Critical milestone (CSF Level 3)**	**Accountable for CSF Level 3 delivery**	**Priority (within scope area)**
<insert CSF Level 1> *<insert specific success criteria>*	*<insert the next level of scope which is required to deliver CSF1>*	*<insert the next level of scope which is required to deliver CSF2>*	*<who in the team is responsible for CSF3>*	*<how important is this objective versus other objectives — need to prioritize>*
	<note that a CSF1 may have a number of associated CSF2s>	*<note that a CSF2 would usually have a limited number of associated CSF3s>*	*<insert name>*	*<insert priority>*
<insert CSF Level 1>	*<insert CSF Level 2>*	*<insert CSF Level 3>*	*<insert name>*	*<insert priority>*
	<insert CSF Level 2>	*<insert CSF Level 3>*	*<insert name>*	*<insert priority>*
	<insert CSF Level 2>	*<insert CSF Level 3>*	*<insert name>*	*<insert priority>*

Critical path of success

A critical success factor (CSF) is an identifiable action/activity that is quantifiable/measurable. It is 'critical' because it has the potential to impact the overall success of the project or sub-project. In order to achieve overall project success each high-level CSF can be joined together to form a critical path. This is known as a critical path of success (Figure 4-5). In order to develop this path a succinct vision of project success needs to be developed. Note that the intake of CSFs does not imply a time based relationship.

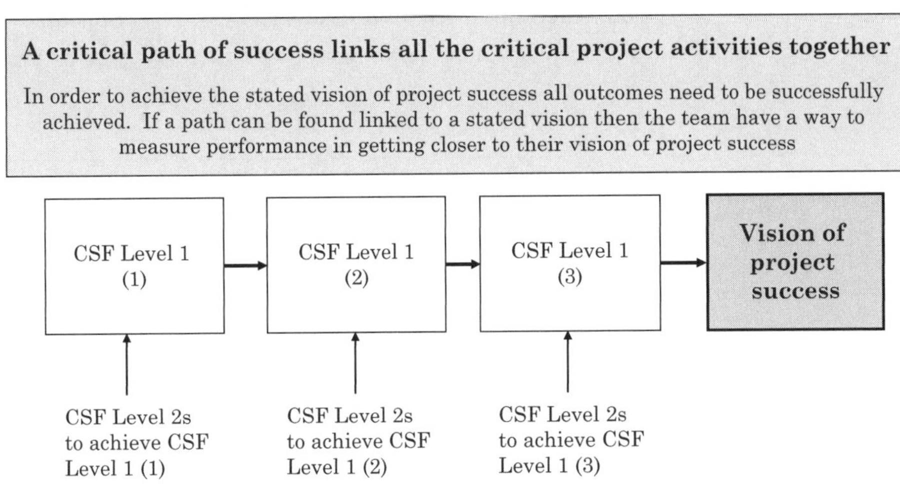

Figure 4-5 The critical path of success

Critical success factor definition

Critical success factors can be determined at a number of levels linked to the scope of the project: in larger projects the activities may be analysed down to Level 4 and in smaller projects or those that are strategic in nature they often only go down to Level 2 and can only be milestone tracked. Typically three levels of CSFs are expected (Figure 4-8).

Scope area (CSF Level 1/CSF1)

In order to successfully complete the project the high-level CSFs need to be defined. These may be descriptions of distinct areas of sub-project scope, for example to successfully sell a new house the path of CSFs at Level 1 in Figure 4-6 is appropriate for the developer:

- It is critical that the design follows the customer parameters otherwise he won't want it.
- It is critical that the construction meets customer and developer expectations.
- Without an effective handover the customer won't buy the house.

Each CSF should be challenged with 'if I don't do this can I still achieve my vision?'

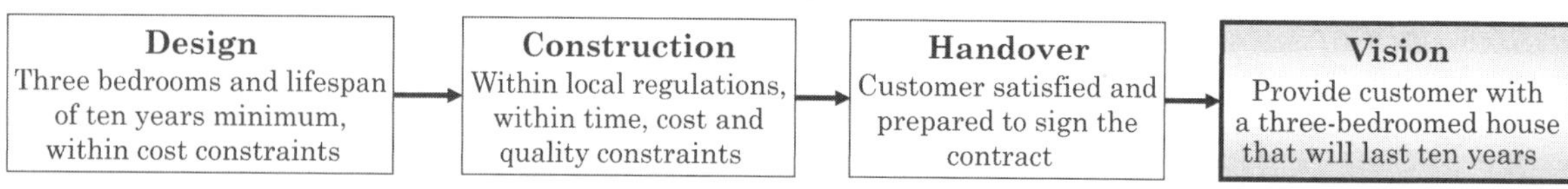

Figure 4-6 An example critical path of success — CSF Level 1

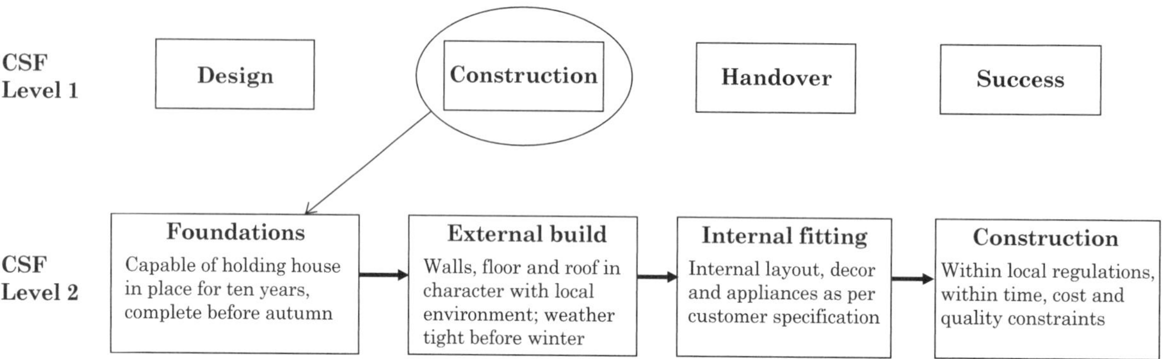

Figure 4-7 An example critical path of success — CSF Level 2

Objective tracking metric (CSF Level 2/CSF2)

In order for each activity (CSF1) to be successful all sub-activities must be successfully completed (CSF2). CSF2 activities tend to be those that can be tracked over time to a clear milestone completion point (Figure 4-7).

Again each of the Level 2 CSFs can be challenged. They should be all critical to the achievement of the specific CSF1.

In Figure 4-7, CSF2(2) the external build, specific objectives can be tracked over time to ensure that they are meeting cost, schedule and scope targets — each day the status of wall and floor build and quality can be reviewed (Table 4-6).

Critical milestone (CSF Level 3/CSF3)

In order for each sub-activity (CSF2) to be successful all associated activities must be successfully completed (CSF Level 3). CSF3s tend to be activities with a clear milestone completion point, for example using the above example of the construction of the external build of a house:

- All walls complete by date (complete means erected and containing any doors and windows and quality checked).
- Roof completed by 'date' (complete means house completely weather tight).

Accountable for CSF Level 3 delivery

It is critical that each CSF3 measured by milestone completion has a named Project Team member who is accountable for its delivery.

Table 4-6 Example Table of Critical Success Factors — building a house

<table>
<tr><th colspan="5">Project Management Toolkit — Table of Critical Success Factors</th></tr>
<tr><td colspan="2">Project: Sell a new house</td><td colspan="3">Date: This year</td></tr>
<tr><th colspan="5">Critical path of success</th></tr>
<tr><td colspan="5">Provide customer with a three-bedroom house that will last ten years</td></tr>
<tr><th colspan="5">Critical success factor definition</th></tr>
<tr><th>Scope area (CSF Level 1)</th><th>Objective tracking metric (CSF Level 2)</th><th>Critical milestone (CSF Level 3)</th><th>Accountable for CSF Level 3 delivery</th><th>Priority (within scope area)</th></tr>
<tr><td rowspan="2">Design
Three bedrooms and a minimum lifespan of ten years, within cost constraints</td><td>House Design
Track design deliverables over 3-month design period</td><td>➤ Design — for approval
➤ Customer approval
➤ Design — for construction</td><td>Architect</td><td>High
Customer has very specific internal design needs</td></tr>
<tr><td>Planning and building Authorizations
Track strategy over 2-month period</td><td>➤ Approval in principle
➤ Final approval</td><td>Project Manager</td><td>Highest
Local conservation area — permission will be difficult</td></tr>
<tr><td rowspan="3">Construction
Within local regulations, within time, cost and quality constraints</td><td>Foundations
Track progress (cost and schedule)</td><td>➤ Excavation
➤ Piling
➤ Ground slab</td><td>Construction Manager</td><td>High
Only requires approval in principle</td></tr>
<tr><td>External build
Track progress (cost and schedule)</td><td>➤ Walls
➤ Floor
➤ Roof</td><td>Construction Manager</td><td>Highest
Requires formal approval</td></tr>
<tr><td>Internal fitting
Track progress (cost and schedule)</td><td>➤ Internal walls
➤ Internal decor
➤ Appliances</td><td>Construction Manager</td><td></td></tr>
<tr><td rowspan="2">Handover
Customer satisfied and prepared to sign the contract</td><td>Customer review</td><td>➤ Customer walk round
➤ Snag items completed</td><td>Construction Manager</td><td></td></tr>
<tr><td>Contract signing</td><td>➤ Contract signed</td><td>Project Manager</td><td></td></tr>
</table>

Priority (within scope area)

Within each CSF Level 1 the corresponding CSF Level 2 and CSF Level 3 should be prioritized. This is to support appropriate resource allocation as necessary rather than choose which CSF needs to be completed. The fact that an item of scope is a CSF means that it has to be completed if the project is to be successful.

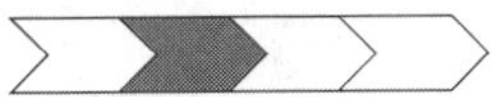

Using the Table of Critical Success Factors

Often it is easier to start development of the table of CSFs on a flip chart using the 'triangle' notation as described in Figure 4-8.

The Project Team should work with the Project Manager to build a detailed hierarchy of CSFs so that the lower levels align completely with the overall project scope. In this way the CSFs can be built from the scope down through the levels and then checked back the other way to confirm complete alignment answering the following questions:

- If this isn't done can this be done? (To go from level to level)
- Can this be done without any other activities? (To check that all critical scope has been identified)

Building the scope in this way incorporates the expertise of the team and allows areas of sub-scope to be packaged up and given to sub-teams for further development whilst fully ensuring that the sub-scope aligns to the overall scope needs of the project.

Once the data is transferred from the flip chart to the table it effectively becomes a work-breakdown structure to be tracked either by the tools suggested (Table of CSFs tracking table) or via the project programme. Note that CSF achievement is not just a time-based success criterion and if using the latter method to track then quality/quantity and functionality should also be checked. Complex CSFs may be tracked through the use of more complex progress and performance measurement tools, some examples of which are introduced in Chapter 5.

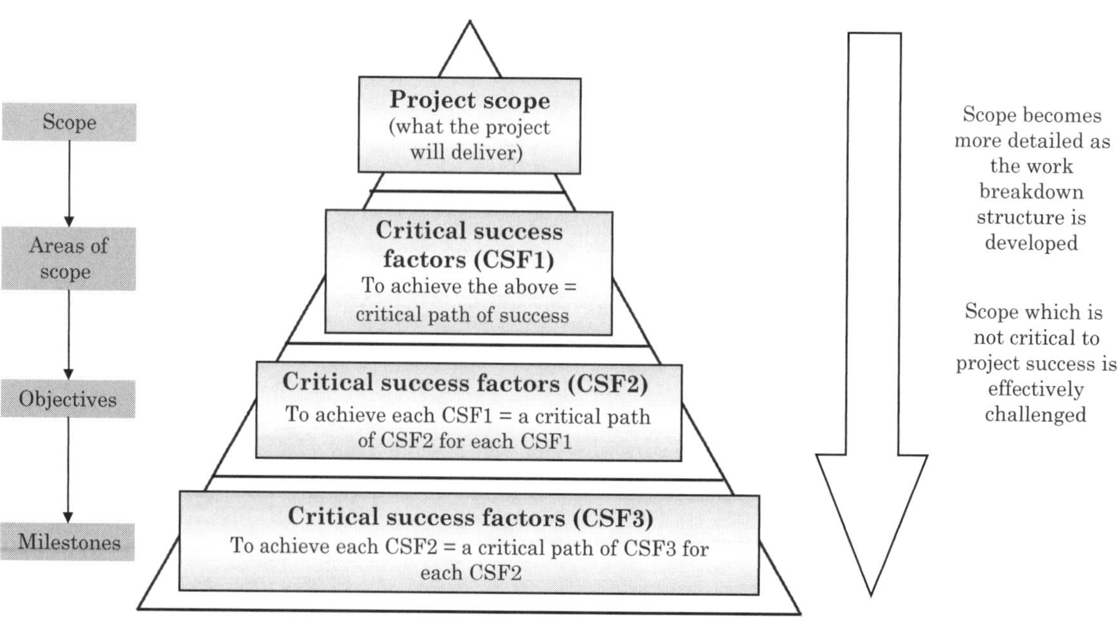

Figure 4-8 A hierarchy of critical success factors

Tool: RACI Chart

The aim of the RACI Chart (RACI is defined in Table 4-7) is to support the development of an appropriate Project Team organization through the development of team roles and responsibilities.

Depending on the activity within the project it may be appropriate to consider the roles and responsibilities of key stakeholders such as the sponsor and customer representatives, who are a part of the wider project organization.

The RACI Chart explained

The tool contains three sections:

- Activities.
- Names.
- RACI allocation.

Table 4-7 shows the RACI Chart with high-level guidance on how to complete each section. The following are additional notes to support tool completion.

Activities

The list of activities will depend on the ultimate use of the RACI Chart for example:

- A RACI Chart for inclusion in a project delivery plan would need to list all the high-level project management activities.
- A RACI Chart for the development and management of a sub-project would contain more detailed activities as related to that sub-project.

Table 4-7 The RACI Chart explained

Project Management Toolkit — RACI Chart				
Project: *<insert project title>*		**Date:** *<insert date>*		
Names → Activity ↓	***<insert Project Team member name>***	***<insert Project Team member name>***	***<insert stakeholder name>***	***<insert stakeholder name>***
<insert activity>	R	A	C	I
<insert activity>	Responsible	Accountable	Consulted	Informed
<insert activity>	*<decide who will complete each activity>*	*<decide who is accountable for the activity completion>*	*<decide who needs to be consulted for the activity to be completed>*	*<decide who needs to be informed of the results>*

Names

The list of names will depend on the ultimate use of the RACI Chart, for example:

- A RACI Chart for inclusion in a project delivery plan would need to list all key team members and key project stakeholders.
- A RACI Chart for the development and management of a sub-project would contain specific team members assigned to that sub-project.

RACI Allocation

For each activity the table needs to contain at least one 'A' and usually one 'R' with as many 'C's or 'I's as is necessary.

- R = Responsible — usually only one person can take responsibility for completing an activity. Where this becomes difficult it would usually indicate that the activity listing needs to be more detailed or an additional RACI Chart developed for a sub-project.
- A = Accountable — only one person can be held accountable for an activity. This is where the 'buck stops'. In terms of overall benefits delivery the sponsor would be an A; whereas the Project Manager is an A in terms of overall project delivery.
- C = Consulted — the number of people who can support the completion of an activity will depend on that activity and the input required. For example a Project Manager would consult with the sponsor and the Project Team whilst developing the project delivery plan.
- I = Informed — the number of people who need to be kept informed of the progress of the activity will depend on how the information would be used to progress other project activities or as a part of the Stakeholder Management Plan (see page 64). For example a summary project progress report would be completed by the Project Manager (AR) with data provided by the Project Team (C) and issued to the sponsor (I).

Using the RACI Chart

The RACI Chart is one method to support the effective articulation of team roles and responsibilities. The following are needed before the chart can be used effectively:

- An understanding of the project scope such as the draft development of the work breakdown structure.
- A view of the likely project organization (Figure 4-9).
- A view on the likely mix of skills needed within that organization (Figure 4-10 which shows a method to select appropriately skilled core team members).

There are many techniques for the development of a RACI Chart however often the most effective is based around a Project Team exercise using the selected team members:

- A blank copy of a RACI Chart should be printed as large as possible (or alternatively drawn by hand on a flip chart).
- The names and roles sections should be populated based on the project organization leaving a few blank columns:
  - Acknowledge that additional roles needed within the organization may need to be identified.
  - Remember to put up key external team members.

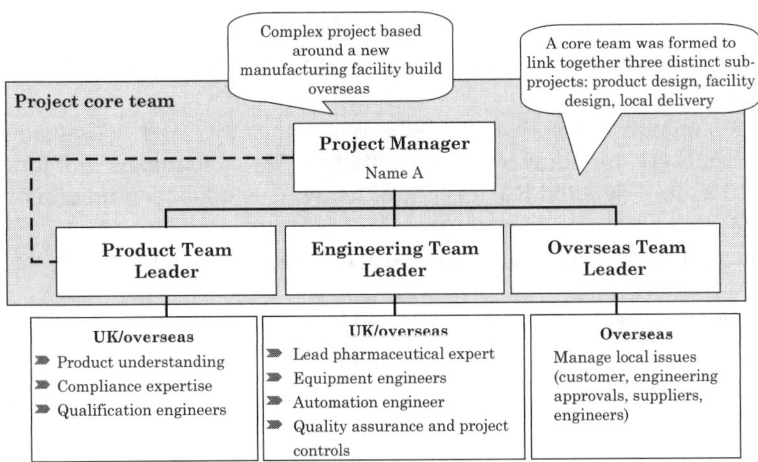

Figure 4-9 Example — project organization

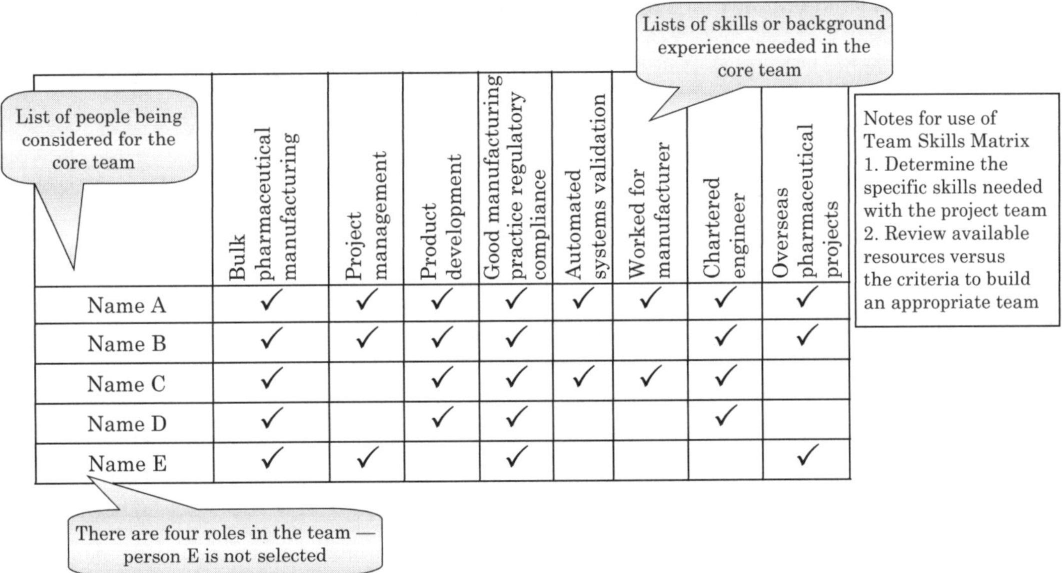

	Bulk pharmaceutical manufacturing	Project management	Product development	Good manufacturing practice regulatory compliance	Automated systems validation	Worked for manufacturer	Chartered engineer	Overseas pharmaceutical projects
Name A	✓	✓	✓	✓	✓	✓	✓	✓
Name B	✓	✓	✓	✓			✓	✓
Name C	✓		✓	✓	✓	✓	✓	
Name D	✓		✓	✓			✓	
Name E	✓	✓		✓				✓

Figure 4-10 Example — understanding the required team skill mix

- Based on current understanding of the scope start to populate the activities section leaving many blank rows:
  - Acknowledge that additional activities may be identified which are not in the current scope (these are likely to be additional detail to critical scope already identified).
  - Remember to insert key external activities.

- The team should then systematically go through each activity and ask:
  - Where does the 'buck stop' with this activity? — Who gets into trouble if it doesn't happen? (Assigning the A)
  - Who actually completes this activity? — Who really does the work? (Assigning the R — remember that if there are many Rs for an activity it probably means that it needs to be developed into a more detailed RACI Chart of its own, as it is made up of more separate activities)
  - Who can help get this activity completed? — Who provides valuable data or feedback? (Assigning the C)
  - Who needs to know that this has happened and how? (Assigning the I)
- The group should be asked to challenge the Cs and Is:
  - It is usual for a team to think that they have to consult with more people than they really will and also that more people need to be informed than actually do.
  - The group should be reminded that the assigning of C and I is related to just that specific activity.
- The group should be asked to challenge whether everyone associated with an activity has been identified:
  - Review versus stakeholder plans and external supply plans.

In this way the team gains 'buy-in' to their roles and responsibilities that are key to project success.

Tool: Control Specification Table

The aim of this tool is to support the development of robust control strategy by assessing the appropriate tools and techniques that should be used on a specific project. Elements of a control strategy include but are not limited to:

- Cost control.
- Schedule control.
- Scope control.
- Change control.
- Progress measurement.

Table 4-8 The Control Specification Table explained

***Project Management Toolkit* — Control Specification Table**		
Project: *<insert project title>*	**Date:** *<insert date>*	
Cost control		
Cost objective/CSF	**Control methodology**	**Responsibility**
<list the appropriate project objectives/CSFs>	*<outline the control techniques or processes to be used — include change control and progress measurement>*	*<insert the name of the person who will manage cost control>*
Schedule control		
Schedule objective/CSF	**Control methodology**	**Accountability**
<list the appropriate project objectives/CSFs>	*<outline the control techniques or processes to be used — include change control and progress measurement>*	*<insert the name of the person who will manage schedule control>*
Scope control		
Scope objective/CSF	**Control methodology**	**Accountability**
<list the appropriate project objectives/CSFs>	*<outline the control techniques or processes to be used — include change control and progress measurement>*	*<insert the name of the person who will manage scope control>*

The Control Specification Table explained

The tool contains three sections:

- Cost control.
- Schedule control.
- Scope control.

Table 4-8 shows the Control Specification Table with high-level guidance on how to complete each section. The following are additional notes to support tool completion.

All three elements of a control strategy should consider the appropriate control methodology and tools, how progress will be measured and how change is appropriately managed.

Cost control

In order to control costs during the delivery of a project a robust cost estimate must be developed. Typically projects involving capital expenditure would require these for formal funding approval however cost planning within revenue-based projects is not as common. Cost planning involves:

- The set-up of a baseline cost report through costing the activities required by the project scope, that is to say it is linked to the work breakdown structure.
- A forecast cash-flow analysis.
- A cost contingency management plan — challenging the cost plan through identification of cost risks.

It is not the intent of this book to expand on any particular cost estimate technique as there are many texts readily available to support this activity. Most will claim a cost accuracy based on the level of scope developed. Examples include vendor quotations, historic data, factorial costs linked to major items or a building floor area, and complete materials listing.

Table 4-9 demonstrates how the basic principles of cost planning can be developed into a typical cost plan report.

A cost plan should be able to confirm to the business the capital and revenue funds needed to deliver the required project; it therefore forms the baseline from which to control costs and is easily converted into a cost report.

At this stage note that cost contingency is the amount of funds allocated per item based on the risk that the cost will exceed the estimate for any number of reasons, the main one is estimate accuracy. Cost contingency should be money which it is assumed will be spent and so it should form a part of the funding requests.

Additionally for projects which are to extend over a longer duration some allowance for cost escalation may also need to be made — to allow for price increases over the duration of the project.

Schedule control

In order to control the schedule during the delivery of a project a robust schedule must be developed at the planning stage. Typically projects involving capital expenditure would require these for formal funding approval, however, schedule planning within revenue based projects is usually not as detailed. Schedule planning involves:

- The development of a baseline logic linked schedule based on estimating the duration and interdependency of the activities required by the project scope — it is linked to the work breakdown structure.

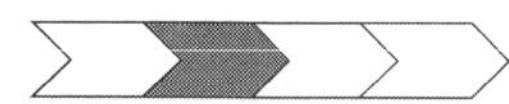

Table 4-9 A typical cost plan report

Item in cost plan	Cost estimated	Method of estimation	Accuracy of estimate	Cost contingency allocated	Cash flow profile
Capital equipment	$65000	Vendor quotation	±10%	±10%	Month 1 — $10 000 Month 3 — $15 000 Month 6 — $15 000 Month 8 — $20 000 Month 9 — $5 000
Capital installation	$12 000	Factored from main equipment costs	±20%	±20%	Month 10–12, $4 000 monthly payments
Total capital	**77 000**			*<insert total required cost contingency>*	**Convert to a cash flow trend**
Project management man-hours	140 days × $day rate	Daily rate at average of 50% of time on project	±5%	Allow an additional week full-time at the end	Month 1–14 equal amounts
Materials to support commissioning	$10 000	Purchase price × amount	Fixed	None	Month 13 one payment
Total revenue	*<insert total revenue>*			*<insert total required cost contingency>*	**Convert to a cash flow trend**

- A review of the critical path.
- Challenging the schedule based on resource constraints and through identification of schedule risks.

It is not the intent of this book to expand on any particular schedule development technique or software tool as there are many texts readily available to support this activity. However some overview principles are presented:

Critical path analysis

The critical path through a project is defined as the longest chain of dependent steps and is developed from a consideration of time. For any project an understanding of the critical path is crucial. Note that the duration for any activity will depend on the level of resources available to that activity (capacity and capability) and so the critical path will be impacted by decisions regarding Project Team size and skills profile.

Presenting schedule information

The Gantt Chart or Bar Chart is possibly the most common form of presenting a schedule. This is a time-based bar chart and it is usual to link related activities together (Figure 4-11).

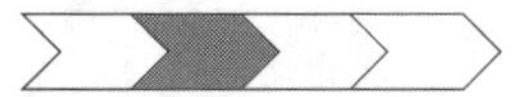

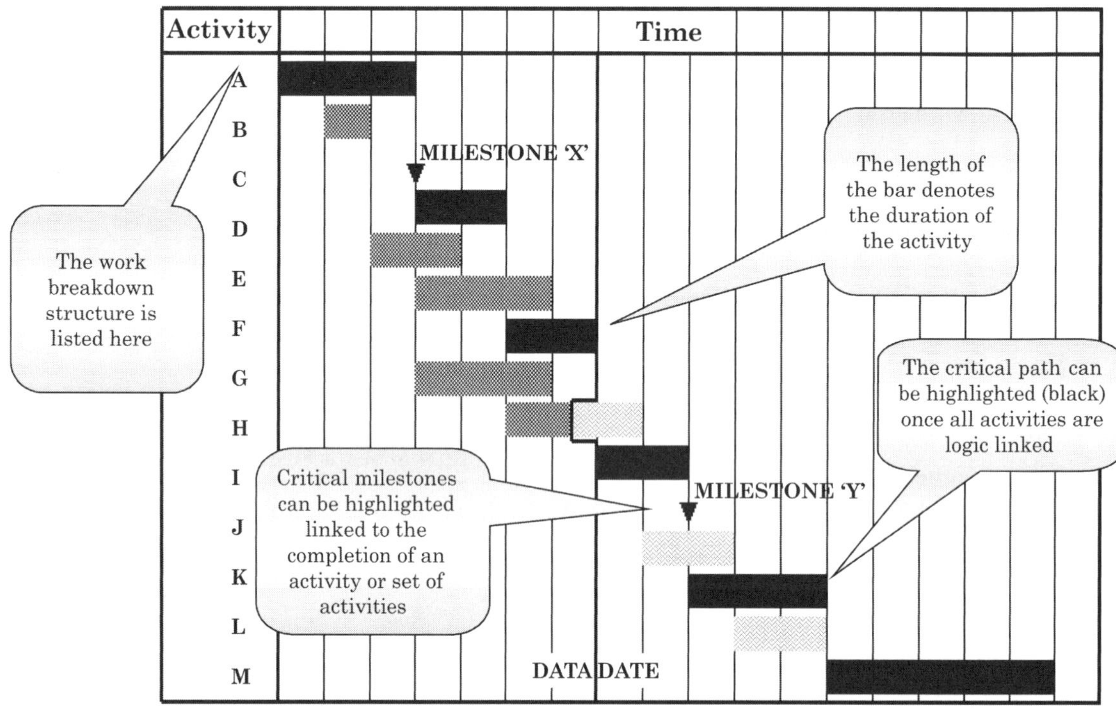

Figure 4-11 An example Gantt Chart

The PERT Chart or Precedence Network is a way of building and demonstrating the logic links or relationships within a group of activities.

The Milestone Chart is a method to summarize key dates in a project. For some simple and short duration projects this may be the main schedule control tool. It can also be used to check and challenge the Table of Critical Success Factors (see page 70), which develops critical milestones from a project perspective.

Activity Plans are detailed action plans which may be used within sub-projects.

Scope control

In order to control scope during the delivery of a project a robust scope definition must be developed at the planning stage. This must consider:

- Quantity.
- Quality.
- Functionality.

Scope can be controlled through the development of key deliverables, with assigned quality and functionality, and linked to key project decisions. All the activities developed via consideration of the critical path of success (see page 71), which form the work breakdown structure, should support the delivery of the project scope.

The biggest issue which impacts scope control is change. Change of scope will impact both cost and schedule as well as the project's ability to meet the customer's needs. A robust baseline scope defined via activities, deliverables and decisions is required in order to analyse any requested change, which occurs once the project commences delivery.

Using the Control Specification Table

Control is at the core of effective project delivery and so the development of an appropriate control strategy is crucial for any Project Manager. The Control Specification Table is the culmination of a series of planning activities and decisions and is based on the robust development of scope in particular but clearly linked to many of the planning activities outlined in the 'How?' Checklist (see page 56).

Depending on the project size, the Control Specification Table could be completed by the Project Manager in isolation or with a sub-set of the Project Team.

However once the table is developed, the results need to be shared with the team members who will initially support the development of the baseline cost, schedule and scope.

Cost

The cost estimation activities must be delegated to the appropriate team member who will be accountable for that element. Generating team buy-in to cost targets and contingency management methodologies is crucial.

Schedule

There are many techniques for the development of a project schedule however the most effective is based around a Project Team exercise:

- The team should be asked to write the individual elements of the work breakdown structure on Post-it® notes (initially starting at CSF Level 1 but working down into detailed tasks).
- The team should be asked to consider the nature of the dependencies between the CSFs — note that this may not be a time based dependency, it may be a quality and/or time dependency, for example if the installation of the reactor doesn't meet quality standards then the operations team cannot commence commissioning.
- The team should be asked to note durations for each activity and to include any assumptions, for example an activity may only take 1 week using two senior engineers but would take 2 weeks using one full-time junior engineer (assuming he is 50% efficient in comparison to the senior engineer).
- The schedule should be converted to the most appropriate form for that project (note that for some projects it can be left as a 'live' schedule on the wall of a central project area).

In this way the team gain 'buy-in' to the schedule target which they must meet if the project is to be successful.

Scope

Scope development via the Table of Critical Success Factors has already been described (see page 70).

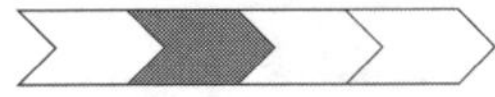

Case study D — if only the 'How?' Checklist had been around then!

This case study is based on a real example of a project which went 'out of control' almost as soon as it started. The 'How?' Checklist is used here to trouble shoot — 'how was the project planned to be delivered in control?'

Blender upgrade

Situation

The project was a very simple one with only one sub-contractor with a single task to complete. The manufacturing manager requested that the site project department complete the project. The scope of the project was to increase the volume of a large capacity solids blending machine for a well-established manufacturing facility, with the works to be carried out during the annual shutdown. The work required the removal of the existing top of the blender, welding of a new section into the wall and then welding of the original top section to the new section. The project engineer selected this method. The time requirement for the contractor's work had been estimated from information gathered from the original sales visit to the site, and had been made by the contractor's sales representative. The annual shutdown period was well in excess of the time requirement to complete the works. The manufacturing department had their own group of engineers. Because of the simple nature of the project the scope of the works was not discussed with the manufacturing engineers and only the manufacturing manager was aware of the scope requirements and that the works were to be carried out. No project schedule was developed because the works appeared easily achievable within the requirements of the annual shutdown.

The project

The project went out of control almost immediately that the contractor arrived on site. Neither the manufacturing manager or the project engineer had informed the plant supervisors that permits to work and electrical equipment isolations would be required as both had assumed that the other would do it. Therefore the contractor did not start on the morning of their first day. It became apparent within a few days that the scope of works would overrun the original time estimates as the time estimate had been developed by the contractor's sales representative and not by engineering personnel. Initial reassessments of the new time requirements were optimistic as the time available within the annual shutdown appeared sufficient. Following this, a realization was made that the contractor's and the project engineer's understanding of the scope of works were not the same. On completion of the welding the welds had to be polished to meet the original finish of the blender, whereas the contractor only planned to grind them flat. This added a further time requirement to the project and had a cost implication.

The outcome

The project was not completed within the time requirements of the annual shutdown and prevented the entire plant from restarting. Production was lost and customer orders were missed.

Because the project was simple and as initial expected time requirements were well within the time requirements for the annual shutdown the impact of the works not being completed on time had not been reviewed.

At project close-out the manufacturing department's engineers stated that once the works had commenced and they were aware of the scope they were sure that the work could not be completed within the annual shutdown requirements. They also made suggestions as to a better method by which the works could have been completed with more of the required works carried out before the shutdown thus requiring a shorter time on site. They also suggested that had they had an opportunity to challenge the scope and the plan, the project may have had a much better outcome.

The project was not treated like a project because of its simplicity.

There was a lack of communication between project engineers and stakeholders (Table 4-10).

Table 4-10 A completed 'How?' Checklist — case study D

Project Management Toolkit — 'How?' Checklist	
Project: Blender volume increase	**Project Manager:** Confidential
Date: Confidential	**Page:** 1 of 4
Stage One check	
Have there been any changes since Stage One completion? (Development of the business case and project kick-off may be some time apart) No project plan was ever developed and this checklist has been completed as part of a trouble shooting exercise following completion of the project. The basic scope requirement for the project did not change as this was a very simple project	
Sponsorship	
Who is the sponsor? (The person who is accountable for the delivery of the business benefits) No sponsor was formally identified as the project was considered relatively simple and of short duration **Has the sponsor developed a communication plan?** No — communication was left to the Project Team	
Benefits management	
Has a benefits realization plan been developed? The Project Team did not consider the impact of the project on the business — either positive or negative (which is how it turned out) **How will benefits be tracked? (Have they been adequately defined?)** The manufacturing team routinely track capacity	
Business change management	
How will the business change issues be managed during the implementation of the project? (Are there any specific resources or organizational issues?) The project had no apparent impact on the business **Have all project stakeholders been identified? (Review the stakeholder map from Stage One)** A full list of key stakeholders was not developed. The Project Engineer had only informed the Manufacturing Manager assuming that he would inform all other necessary stakeholders within manufacturing. The Manufacturing Manager assumed that the Project Engineer was communicating the project requirements to stakeholders within the manufacturing department **What is the strategy for handover of the project to the business? (Link this to the project objectives)** None	

(continued)

Table 4-10 (Continued)

<table>
<tr><th colspan="2">Project Management Toolkit — 'How?' Checklist</th></tr>
<tr><td>Project: Blender volume increase</td><td>Project Manager: Confidential</td></tr>
<tr><td>Date: Confidential</td><td>Page: 2 of 4</td></tr>
<tr><th colspan="2">Scope definition</th></tr>
<tr><td colspan="2">Has the scope changed since Stage One completion?
The development of the design did reveal an additional problem to be solved within the scope which had links into both safety and personnel areas. This requirement was the key driver for selection of how the scope would be completed — to remove the blender top, weld in a new section and refit the top section. Selection of this method maintained the status quo and did not require the project to get involved in the two problem areas. Improvement may have been possible but this had not been within the scope of the project
Have the project objectives been defined and prioritized? (What is the project delivering?)
➤ Increase the blender volume to prevent damage
➤ Ensure that the changes meet health and safety requirements
➤ Ensure that the changes do not cause additional operational requirements and that blender lids close and interlock as required</td></tr>
<tr><th colspan="2">Project type</th></tr>
<tr><td colspan="2">What type of project is to be delivered? (For example engineering or business change)
This is an engineering capital project improving health and safety and operational ability issues within an existing plant
What project stages/stage gates will be used? (Key milestones, for example funding approval which might be go/no go points for the project)
This was a very simple project and a plan was not produced, therefore the setting of milestones was not considered. The plan was therefore only verbal and was not communicated to key stakeholders such as the manufacturing engineers and supervisors only the manufacturing manager. This lack of communication meant that there was no challenge to the plan which would have revealed its shortcomings and also would have produced a much better proposal for completion of the project scope</td></tr>
<tr><th colspan="2">Funding strategy and finance management</th></tr>
<tr><td colspan="2">Has a funding strategy been defined? (How will the project be funded and when do funds need to be requested?)
Funding for the project was achieved through the normal site small projects approval route and was identified in annual capital plan for completion within the defined year
How will finance be managed?
The project was managed by the site project department. A project is permitted to overspend by up to 10% without project review and re-approval</td></tr>
</table>

(continued)

Table 4-10 (Continued)

<table>
<tr><th colspan="2">Project Management Toolkit — 'How?' Checklist</th></tr>
<tr><td>Project: Blender volume increase</td><td>Project Manager: Confidential</td></tr>
<tr><td>Date: Confidential</td><td>Page: 3 of 4</td></tr>
<tr><th colspan="2">Risk and issue management</th></tr>
<tr><td colspan="2">Have the CSFs changed since Stage One completion? (As linked to the prioritized project objectives and the critical path through the project risks)
An additional requirement to consider health and safety and operations personnel requirements was requested
Have all project risks been defined and analysed? (What will stop the achievement of success?)
No consideration was given to project risk. The main requirement for the project was to be completed within the annual shutdown requirements. As the initial estimates of time were easily achievable within the annual shutdown no analysis of the impact on a time overrun by the project was completed. A review of the project plan was never even completed as no such plan was ever prepared. No copy of a risk assessment was available
What mitigation plans are being put into place?
None
What contingency plans are being reviewed?
None</td></tr>
<tr><th colspan="2">Project organization</th></tr>
<tr><td colspan="2">Who is the Project Manager?
Confidential
Has a project organization for all resources been defined? (Include the Project Team and all key stakeholders)
No organization chart was developed for the project. No formal roles and responsibilities were defined</td></tr>
<tr><th colspan="2">Contract and supplier management</th></tr>
<tr><td colspan="2">Has a strategy for use of external suppliers been defined? (The reasons why an external supplier would need to be used for any part of the scope)
No strategy was developed for the use of an external contractor. The decision to use an external contractor was made by the project department who also selected the chosen contractor. Discussion with manufacturing engineers revealed that they had completed similar works before and knew of contractors with experience of these types of works. These contractors did not even quote for the project works
Is there a process for using an external supplier? (For example selection criteria, contractual arrangements, performance management)
An external contractor was selected by the project department on the basis of lowest cost. A contractor sales representative had developed the contractor proposal without input from technical staff</td></tr>
</table>

(continued)

Table 4-10 (Continued)

Project Management Toolkit — 'How?' Checklist	
Project: Blender volume increase	**Project Manager:** Confidential
Date: Confidential	**Page:** 4 of 4
Projects controls strategy	
Is the control strategy defined? ➠ *Cost control strategy* — the project works required the placing of a single fixed price order which purchasing paid so therefore no strategy was developed ➠ *Schedule strategy* — no plan was ever completed ➠ *Change control* — change control was not implemented as the method selected for completion of the project ensured that the project did not involve itself in areas which may have required a project change. Consideration of project changes could have produced a better outcome for the project, however possible alternatives were never considered ➠ *Action/progress management* — none, and initial reviews of progress and re-estimation of time requirements were not given proper consideration and did not include the manufacturing engineers as the works appeared to be possible within the requirements of the annual shutdown ➠ *Reporting* — none, the Project Engineer did not report to the annual shutdown coordination group and only informed them of project progress on a need to know basis	
Project review strategy	
Is the review strategy defined? (How will performance be managed and monitored — both formal and informal reviews and those within and independent to the team?) No plan was produced and therefore no review process was required or carried out. The completion of this checklist is the first formal review	
Stage Two decision	
Should the project be progressed further? (Is the project delivery strategy robust enough for project delivery to commence?) Not applicable, the checklist has been completed as part of a troubleshooting exercise following close-out of the project. If this was a live situation the project would not be permitted to proceed based on the above information. Based on this review it is no surprise that the project is over budget and over allotted time. Additionally the missed customer orders caused customer complaints. The relationship between manufacturing engineering and project engineering has been worsened through the implementation of this project	

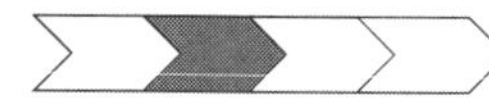

Case study E — using the 'How?' Checklist to re-plan a project

This case study demonstrates what can happen when you use the 'How?' Checklist on a project that is about to commence delivery. As a result the project was slowed down until a robust delivery plan was in place.

Effluent treatment

Situation

The project required the design and construction of an effluent treatment facility on a greenfield site for a medium sized chemicals manufacturer. All design works and civil construction of the new bund to house process equipment was complete. Main site-constructed vessel and pump orders had been placed, mechanical contract tenders had been received as had quotations for construction of new building for electrical and control equipment. It was apparent at this point that the project could not be completed within budget and would be 25–50% more than original estimates. The possibilities for additional funding for the project were investigated and the site advised that no more money could be made available and that the project must be completed within the original budget. At this point the project was stopped and no further orders or construction works were to take place until the project costs could be brought under control. Corporate engineering requested a formal review of the project.

The project

A strategy to focus the project was developed based on the CSF that for the project to be successful it must deliver the functionality shown on the piping and instrumentation diagrams (P&ID). The P&IDs were reviewed and approved by all project stakeholders. Other elements of the project were considered to be more flexible. This focused attention on the critical elements of the project. It was also planned that as little re-design should be completed as possible in order to limit further spending on work that had already been completed.

With this strategy in mind the delivery and site construction of main vessels was allowed to proceed, as these were needed within the requirements of the P&IDs and the order was so well progressed that cancellation of any of the items would be prohibitively expensive.

The P&ID review process identified that duty and standby pumps were not required. The standbys were cancelled and spares which would allow a quick turn around on repair of a faulty item purchased in their place. This strategy was agreed as the impact of a faulty pump was not detrimental to site operations assuming pumps could be repaired within a 24-hour period.

The electrical and instrumentation (E&I) tender package was issued and quotes obtained.

On the basis of the quotes, E&I and mechanical contractors were brought to site to review possibilities for further reducing cost, allowing the site engineers to draw on the contractors previous experiences. The contractors were paid up front to complete a full survey of the design works identifying any required changes and shifting the cost of all further change to the contractor once the contract was placed. Incentives for timely completion of the works, which focused on both contractor and project engineering requirements, were agreed.

The building method for the E&I building was simplified and rather than providing a building with additional spare space as per the site norm, the building was designed to allow easy installation of a second equally sized bay in the future if required. At this point a new project estimate was drawn up making allowances for contingency based on the present project progress — risk of additional cost has been limited for mechanical and E&I contracts.

The outcome

The project was completed slightly under the original budget and was completed to revised schedule requirements (Table 4-11).

Table 4-11 A completed 'How?' Checklist — case study E

Project Management Toolkit — 'How?' Checklist	
Project: Effluent treatment	**Project Manager:** Confidential
Date: Confidential	**Page:** 1 of 4
Stage One check	
Have there been any changes since Stage One completion? (Development of the business case and project kick-off may be some time apart) Project scope has remained fixed, project requirements are defined by completed P&IDs. This checklist has been completed following the stoppage of the project due to budget problems	
Sponsorship	
Who is the sponsor? (The person who is accountable for the delivery of the business benefits) Site Manager **Has the sponsor developed a communication plan?** No	
Benefits management	
Has a benefits realization plan been developed? Not applicable, the project was all about meeting local authority 'consent to discharge' limits which are collected on a frequent basis **How will benefits be tracked? (Have they been adequately defined?)** Not applicable, the main benefit is an operational measure	
Business change management	
How will the business change issues be managed during the implementation of this project? (Are there any specific resources or organizational issues?) Operators are to be trained by manufacturing and technical services, who must also produce operating procedures. The project department is to make design information available as soon as possible **Have all project stakeholders been identified? (Review the stakeholder map from Stage One)** Stakeholders have now been identified and will be included in a review and approval of the P&IDs to ensure that the project will deliver all items necessary for its success **What is the strategy for handover of this project to the business? (Link this to the project objectives)** The project will be handed over to technical services for commissioning, start-up and handover to manufacturing. The project department will be required to provide mechanical and E&I resources during the commissioning period. This should be provided by the installation sub-contractors and will be included in sub-contractors scope of work	

(continued)

Table 4-11 (Continued)

Project Management Toolkit — 'How?' Checklist	
Project: Effluent treatment	**Project Manager:** Confidential
Date: Confidential	**Page:** 2 of 4
Scope definition	
Has the scope changed since Stage One completion? (Has further conceptual design been completed which may have altered the scope?) Whilst project scope has not changed, it is apparent that the original project budget is insufficient to complete the works, and therefore changes must be agreed. The original budget was flawed **Have the project objectives been defined and prioritized? (What is the project delivering?)** The project must deliver a functional effluent treatment plant in order that the site can maintain future consents to discharge. This is as defined by the P&IDs, which will be reviewed and agreed by all stakeholders	
Project type	
What type of project is to be delivered? (For example engineering or business change) Engineering capital project **What project stages/stage gates will be used? (Key milestones for example funding approval which might be go/no go points for the project)** The site construction of main vessels will be allowed to proceed as it has been determined that the order is prohibitively expensive to cancel as it is so well progressed The remainder of the project will not progress until all remaining costs are fixed as far as is possible and a new project budget within fixed quoted costs has been drawn up. This budget must be shown to be robust. An internal site stage gate approval will be introduced to approve the revised budget and plan. Further orders will not be placed until this approval has taken place	
Funding strategy and finance management	
Has a funding strategy been defined? (How will the project be funded and when do funds need to be requested?) The project was approved via normal approval routes and was included on corporate capital plans. An additional funding request has been investigated and the site advised that no increase to funding will be made for this project **How will finance be managed?** The site project department will manage finance with weekly reporting to corporate engineering. No overspend is permitted	

(continued)

Table 4-11 (Continued)

Project Management Toolkit — 'How?' Checklist

Project: Effluent treatment	**Project Manager:** Confidential
Date: Confidential	**Page:** 3 of 4

Risk and issue management

Have the CSFs changed since Stage One completion? (As linked to the prioritized project objectives and the critical path through the project risks)
The critical success factors for the project have been limited to provision of the functionality required by the P&IDs. Project delivery strategy will focus on delivery of this functionality and all other areas of the project will be treated as 'nice to have' and will only be included should sufficient funds be available

Have all project risks been defined and analysed? (What will stop the achievement of success?)
- Change by manufacturing and technical service departments. The possibility for future change is limited by including all stakeholders in the review and approval of the P&IDs. It is also agreed that following approval of the P&IDs a very strict change control procedure will be employed
- Changes to sub-contract requirements due to actual site conditions which are not anticipated within the original design

What mitigation plans are being put into place?
- The installation contractors will be required to complete a survey prior to the start of the works, which will provide them with the opportunity to identify changes. Any further changes required during the construction period will be carried out by the contractor at their cost
- The project will not progress beyond construction of the site vessels until a new budget is prepared and approved. This will include actual quoted costs for all elements of the project

What contingency plans are being reviewed?
A review of construction risks will be completed and mitigation plans considered. This exercise will be completed prior to completion of the new budget including suitable contingency allowances

Project organization

Who is the Project Manager?
Site appointed Project Manager from project department. A new Project Manager will be appointed for the restart

Has a project organization for all resources been defined? (Include the Project Team and all key stakeholders)
Yes, key stakeholders have been identified and informed of the present situation within the project

Contract and supplier management

Has a strategy for use of external suppliers been defined? (The reasons why an external supplier would need to be used for any part of the scope)
Yes, for the restarted project a strategy for the use of external contractors has been developed in line with the notes above. In addition to these requirements, timely completion of the site installation works will be incentivized in order to limit the requirements for site establishment, which could result in additional costs to the project

Is there a process for using an external supplier? (For example selection criteria, contractual arrangements, performance management)
For the restarted project, contractors with a longer term relationship to the site will be utilized. The situation within the project will be discussed with the contractors and they will be requested to make proposals for additional cost cutting in order that the site may benefit from their experience

(continued)

Table 4-11 (Continued)

Project Management Toolkit — 'How?' Checklist	
Project: Effluent treatment	**Project Manager:** Confidential
Date: Confidential	**Page:** 4 of 4
Project controls strategy	
Is the control strategy defined? For the restarted project the following have been defined: ➤ *Cost control strategy* — orders will not be placed until costs are known and fixed ➤ *Schedule strategy* — site installation works will not commence until costs are fixed, scope of work will be surveyed by contractors and time requirements agreed by site and contractor ➤ *Change control* — strict change control procedures will be employed, with budget holders approval required for any legitimate change, in order to identify funding for any such change ➤ *Action/progress management* — installation contract progress will be measured on a weekly basis ➤ *Reporting* — Project Manager will report to the site manager and corporate engineering on a weekly basis	
Project review strategy	
Is the review strategy defined? (How will performance be managed and monitored — both formal and informal reviews and those within and independent to the team?) Project will be reviewed and reported weekly. Changes to project requirements must be approved by budget holder. Increase to overall project cost will not be allowed	
Stage Two decision	
Should the project be progressed further? (Is the project delivery strategy robust enough for project delivery to commence?) Project will not commence until a new and 'within cost' project budget is prepared and approved. As a result of this review a revised delivery plan has been developed for the restarted project. Before restart this will be issued to all senior stakeholders for approval	

Case study F — using the how tools to plan a project

This case study uses all of the Stage Two tools and is based on a real example in industry. The feedback at the end of the project was that 'initially I was worried because you didn't seem to be doing anything — I can now see that the planning has supported a successful outcome'.

Relocation project

Situation

An organization was building a new headquarters in order to better integrate key business operations, which are currently located around several small offices and one other larger site. The new building project is well established and progressing to plan and the executive board want the relocation project to be developed — delivery of the actual moves of all business units from several locations to their optimum location. The board expects financial benefits from the closure of the smaller offices and from the high utilization of the new building and any remaining sites.

The project

The purpose of the project was to evaluate the optimum location options for the organization and then to successfully deliver the approved solution in compliance with business needs. Initially the project was centred on the new building project and also the facilities management needs of each of the existing locations.

The project director works directly for the sponsor and has appointed an experienced engineering Project Manager so that professional project management practices are used. The first task completed by the Project Manager was the development of an extensive project delivery plan. This enabled the Project Manager to talk to all of the project stakeholders and get a good overall view of the current situation prior to making any key decisions. Table 4-12 represents a 'snapshot' of the draft work which was completed during the planning phase. It highlights that the 'softer' people, culture and business change issues had not been effectively integrated into the project scope.

By the time project delivery commenced the project was effectively managed and supported by a hierarchy of project delivery plans all aligning to the overall project delivery plan (Figure 4-12).

The outcome

The project delivery plan proved to be the core document which kept a geographically and culturally diverse team focused.

The plan also fundamentally changed the nature of the project:

- Prior to the development of the project delivery plan the project was considered to be a facilities management project.
- During the early stages of the development of the project delivery plan it became clear to the Project Manager and all key stakeholders that this was a business change project based around a facilities project; as a result of this project many people would be working in a different place and in a different way. Delivering this project successfully had the potential to support and increase focus on the culture change initiatives which had been developing independently.

The project delivery plan was used as a 'live' tool throughout the project and supported the ultimate success — delivery of all the required benefits (both 'hard' and 'soft').

Table 4-12 A completed 'How?' Checklist — case study F

<table>
<tr><th colspan="2">Project Management Toolkit — 'How?' Checklist</th></tr>
<tr><td>Project: Relocation project</td><td>Project Manager: Confidential</td></tr>
<tr><td>Date: Confidential</td><td>Page: 1 of 4</td></tr>
<tr><td colspan="2">Stage One check</td></tr>
<tr><td colspan="2">Have there been any changes since Stage One completion? (Development of the business case and project kick-off may be some time apart)
The business case had only considered the financial benefits of the location of business units in fewer locations aligned with increasing the utilization of the new building to its maximum. The remainder of the project plan will also include the people, culture and business change benefits. This is at a very early stage in the project and this fundamental shift will not be a problem. The actual design of the 'solution' has not been completed</td></tr>
<tr><td colspan="2">Sponsorship</td></tr>
<tr><td colspan="2">Who is the sponsor? (The person who is accountable for the delivery of the business benefits)
The Chief Financial Officer (who is a member of the board)
Has the sponsor developed a communication plan?
A draft communication plan has been developed by the Project Manager in conjunction with the Project Director who will review with the sponsor. This is based on the Stakeholder Management Plan (Table 4-13) which identified the importance of early communication with the other business unit executives</td></tr>
<tr><td colspan="2">Benefits management</td></tr>
<tr><td colspan="2">Has a benefits realization plan been developed?
No — the Benefits Specification Table was completely redeveloped in line with the development of both 'hard' and 'soft' benefits. Once this is fully approved a realization plan will be developed and linked to the overall project schedule
How will benefits be tracked? (Have they been adequately defined?)
The benefits have now been realigned into three basic categories:
➤ Financial benefits associated with the optimum use of the companies' assets
➤ Business unit efficiency benefits associated with the co-location of business units
➤ Cultural benefits associated with the development of different ways of working
Additionally the final solution will need to consider minimizing the total number of moves (people and cost issue) and the future flexibility of the option (future optimum use of assets). In other words, for example, if people have to move again in a years time the project has not been successful</td></tr>
</table>

(continued)

Table 4-12 (Continued)

Project Management Toolkit — 'How?' Checklist	
Project: Relocation project	**Project Manager:** Confidential
Date: Confidential	**Page:** 2 of 4
Business change management	
How will the business change issues be managed during the implementation of the project? (Are there any specific resources or organizational issues?) The business change aspects were included within the scope of this project via the development of location sub-projects and business unit sub-projects. In this way the business changes were aligned and focused on the delivery of the overall benefits — the initial concern was that each individual business unit could have diverse needs and drivers and that this needed coordination **Have all project stakeholders been identified? (Review the stakeholder map from Stage One)** The stakeholder map was large and complex and included Senior Board Members, Site Directors, Site Facility Engineers and end users (people who would actually have to move and work differently). Table 4-13 is an extract from the completed Stakeholder Management Plan **What is the strategy for handover of the project to the business? (Link this to the project objectives)** The handover milestones were incorporated into the location plans as handover was determined to be after the physical relocation had occurred	
Scope definition	
Has the scope changed since Stage One completion? (Has further conceptual design been completed which may have altered the scope?) The scope has been clarified further and now explicitly includes various types of projects: construction (the new building), engineering (location revamp), relocation (moving whole business units); business change (how the business units and people within them will be expected to work as a result of the move). Additional scope relating to various functional needs have also been identified, for example human resources, communications **Have the project objectives been defined and prioritized? (What is the project delivering?)** The prioritized objectives have now been clarified and are summarized in Table 4-14	
Project type	
What type of project is to be delivered? (For example engineering or business change) Overall this is a business change project with many types of sub-projects (for example engineering, relocation, human resources) **What project stages/stage gates will be used (Key milestones for example funding approval which might be go/no go points for the project)** A specific project roadmap has been developed for this project that uses various stage gates. The next critical stage gate is the approval of the project delivery plan	

(continued)

Table 4-12 (Continued)

Project Management Toolkit — 'How?' Checklist	
Project: Relocation project	**Project Manager:** Confidential
Date: Confidential	**Page:** 3 of 4
Funding strategy and finance management	
Has a funding strategy been defined? (How will the project be funded and when will the funds be requested?) Yes, the funding is complex due to the nature of how the costs can be accounted for within the company — a mixture of revenue and capital and a mixture of overhead and business unit allocation **How will finance be managed?** A functional sub-project has been set up and will be led by a team leader from the corporate finance group. Together this leader and the Project Manager have agreed a cost plan template which will be issued to the location team leaders. The majority of the funds will be spent in actually moving the business units (equipment and people). There is a central fund for the core team which is to be managed by the Project Manager	
Risk and issue management	
Have the CSFs changed since Stage One completion? (As linked to the prioritized project objectives and the critical path through the project risks) Yes, an overall critical path of success has been defined **Have all project risks been defined and analysed? (What will stop the achievement of success?)** Yes, a detailed risk assessment has been conducted for this project and all associated sub-projects **What mitigation plans are being put into place?** The new project organization was developed as a direct response to high (red) risks, as was the decision to use an external communications consultant. Getting better data from the business units has been helped by the sub-team set-up and the use of end user business unit representatives (to get 'softer' data) **What contingency plans are being reviewed?** At the moment no contingency plans are in development — reviewing effectiveness of mitigation plans will determine if more needs to be done at the next review	
Project organization	
Who is the Project Manager? There is a Project Director and a Project Manager each with distinct roles and responsibilities as outlined in the project RACI Chart (Table 4-15) **Has a project organization for all resources been defined? (Include the Project Team and all key stakeholders)** Yes, an extract from the agreed RACI Chart is included (Table 4-15). This is aligned to the Project Team organization which has recently been developed. The organization shows functional, location and business unit team leaders who will manage specific sub-projects	
Contract and supplier management	
Has a strategy for use of external suppliers been defined? (The reasons why an external supplier would need to be used for any part of the scope) Yes, external suppliers will be needed to support communications and also the physical move process **Is there a process for using an external supplier? (For example selection criteria, contractual arrangements, performance management)** Yes, this is in place. Appointment of the communications consultant is imminent	

(continued)

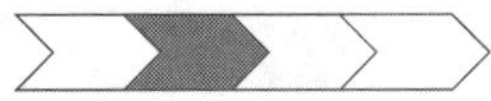

Table 4-12 (Continued)

Project Management Toolkit — 'How?' Checklist	
Project: Relocation project	**Project Manager:** Confidential
Date: Confidential	**Page:** 4 of 4
Project controls strategy	
Is the control strategy defined? Yes (Table 4-16), all the Sub-Project Team Leaders have bought in to the control strategy: ➤ *Cost control strategy* — costs estimated to be within budget ➤ *Schedule strategy* — schedule developed and on track ➤ *Change control* — new change control system in force to deal with lack of business unit data accuracy ➤ *Action/progress management* — core team action log in place ➤ *Reporting* — monthly one-page reports to commence this month	
Project review strategy	
Is the review strategy defined? (How will performance be managed and monitored — both formal and informal reviews and those within and independent to the team?) The plan is to have monthly core team meetings with all Sub-Project Team Leaders present. Additionally external reviews have been organized with the main engineering project management director. All Sub-Project Team Leaders are expected to meet with their teams at least every 2 months	
Stage Two decision	
Should the project be progressed further? (Is the project delivery strategy robust enough for project delivery to commence?) The project delivery plan has been presented to the board and to all business unit executive teams. Approval to proceed has been given on the basis that the completed project delivery plan (an extensive document for a project of this size and complexity) is a robust plan. The project delivery plan document was made into a summary presentation to facilitate ease of communication and therefore understanding	

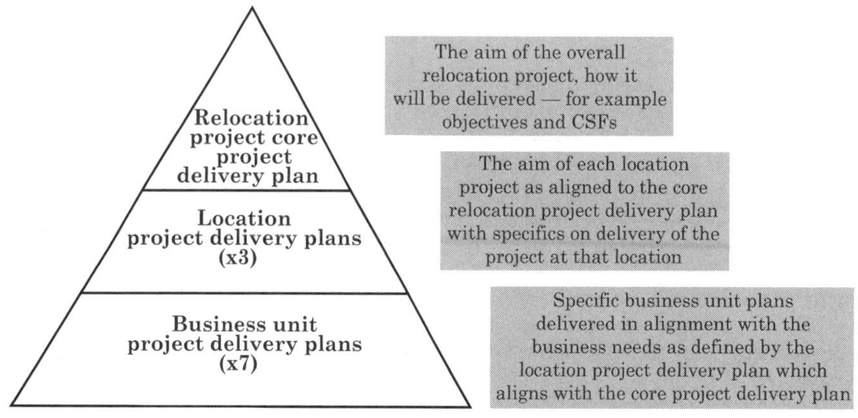

Figure 4-12 Hierarchy of project delivery plans — case study F

Table 4-13 A completed Stakeholder Management Plan — case study F (Note that this is an extract from the plan which is several pages in length). C = champion; S = sponsor; (r) = reinforcing; (a) = authorizing; A = change agent; T = target

Project Management Toolkit **— Stakeholder Management Plan**						
Project: Relocation project				**Date:** Confidential		
Individual stakeholder analysis						
Stakeholder name/role	**Type**	**Current project knowledge**	**Level of influence**	**Current level of engagement**	**Target level of engagement**	**Management of stakeholder**
Name — '*confidential*' (Chief Financial Officer)	S(a)	High	High	High — the benefits from this project are part of the global business plan. He 'opens doors' for the team (into other business units) and can quickly resolve political issues	High — need him to continually remind the other business unit executives of the criticality of this project and for them to release their resources appropriately	Name — '*confidential*' (Project Director) Regular one-to-ones to keep him updated of progress and any issues
Name — '*confidential*' (Site Director — largest location)	S(r)	High	Medium	Medium — he is still treating this a bit like a facility project but as his is the largest site with the biggest engineering impact it is understandable	High — need to spend some time with him to review the non facilities management objectives	Name — '*confidential*' (Project Manager) Infrequent one-to-ones but attend his monthly site management meetings
Name — '*confidential*' (Business Unit Director — largest business unit)	S(r)	Low	High	Low — she is sceptical about the cultural benefits	High — a large number of employees in this business unit will be impacted	Name — '*confidential*' (Business Unit Team Leader)
Name — '*confidential*' (Business Unit Representative)	A	Medium	Medium	Low — trying to release her to be assigned to the Project Team to support local implementation	High — potentially high value in keeping a specific location and a 'problem' business unit up to speed	Name — '*confidential*' (Location Team Leader)
Name — '*confidential*' (Business Unit Representative)	T	Low	Medium	Low — doesn't want to move and has good links in the business unit	Medium — need to manage the negative impact	Name — '*confidential*' (Business Unit Team Leader)

(continued)

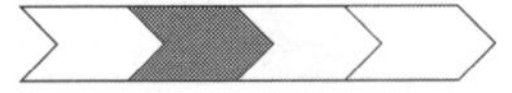

Table 4-13 (Continued)

Project Management Toolkit — Stakeholder Management Plan						
Project: Relocation project			**Date:** Confidential			
Individual stakeholder analysis						
Stakeholder name/role	**Type**	**Current project knowledge**	**Level of influence**	**Current level of engagement**	**Target level of engagement**	**Management of stakeholder**
Name — '*confidential*' (Business Unit Representative)	C, A	High	Low	High — he has been going on about the need for the cultural/ business change issues for some time (from his business unit perspective)	High — his good relationship with the Site Director needs to be put to good use	Name — '*confidential*' (Location Team Leader)
Summary stakeholder analysis						
The organizational breadth of this project has identified that there are a number of very senior stakeholders involved with this project As a result a plan has been developed and is in progress — the project delivery plan has been presented to senior executive teams in all impacted business units and their concerns noted. Additionally the communication plan has been well received (one-page senior executive reports plus newsletter type information to business units via the representatives we have now engaged). Stakeholder management is recognized as a critical part of this project and an employee survey at the end of the project will be used to assess how people (at all levels in the organization) felt as the project progressed						

Table 4-14 A completed Table of Critical Success Factors — case study F

Project Management Toolkit — Table of Critical Success Factors

Project: Relocation project | **Date:** Confidential

Critical path of success

Relocation of a specified number of business units to their optimum location (as defined by asset utilization) within the specified timescale and the specified cost whilst minimizing disruption to normal business operation and employees and maximizing the potential for additional benefits such as operational efficiencies through co-location of business units

Critical success factor definition

Scope area (CSF Level 1)	Objective tracking metric (CSF Level 2)	Critical milestone (CSF Level 3)	Accountable for CSF Level 3 delivery	Priority (within scope area)
Optimal strategy Project delivery plan agreed by board and implemented successfully	Project delivery plan development and use (incorporate business unit data on move needs, track data accuracy and change)	Project delivery plan completion '*date*' Project delivery plan approved '*date*' Project delivery plan review '*date*'	Project Manager	*Highest* The optimal solution is critical to overall project success
	Project expenditure	Cost plan approved '*date*' Capital cost in budget Revenue cost in budget	Financial Functional Leader	*High*
	Project schedule	Master schedule approved '*date*'	Project Manager	*High*
Asset utilization >80% utilization of space in all remaining assets	Asset allocation plan	>90% asset allocation approved '*date*'	Functional Facilities Management Leader	*Highest* Agreement with business unit leaders is critical
	Asset utilization plan	>70% utilization '*date*' >80% utilization '*date*'	Project Leader	*High*
Physical integration >90% business units >100% physically co-located by '*date*'	Master move schedule	Total move completion '*date*'	Functional Facilities Management Leader	*Highest*
		Location move completion '*date*'	Location Team Leader	*Medium*
		Business unit move completion '*date*'	Business Unit Team Leader	*High*
Business changes Minimize disruption to business and maximize cultural integration	Number of interim moves	<10% of total moves involve an interim move	Functional Facilities Management Leader	*Highest*
	Transport issues log	<5% of staff impacted by transport issues	Functional Facilities Management Leader	*Medium*
	Schedule of human resources policies	Human resources policies in place by first move	Functional Human Resources Leader	*Medium*

(continued)

Table 4-14 (Continued)

Project Management Toolkit — Table of Critical Success Factors				
Project: Relocation project		**Date:** Confidential		
Critical path of success				
Critical success factor definition				
Scope area (CSF Level 1)	**Objective tracking metric (CSF Level 2)**	**Critical milestone (CSF Level 3)**	**Accountable for CSF Level 3 delivery**	**Priority (within scope area)**
	Individual human resources issues log	<5% of staff are disrupted by the move	Functional Human Resource Leader	*High*
	Business unit commissioning plan	>95% of staff receive timely and high quality communication	Business Unit Team Leader	*High*
Asset disposal All vacated assets disposed of	Site disposal schedule	Site disposal complete '*date*'	Project Leader	*Highest*
	Site disposal budget	Site disposal release of funds '*date*'	Project Leader	*High*
Asset availability All assets available for occupation	New build plan	Ready for visits date Ready for occupation '*date*'	New build Project Manager	*Highest* Starts the moves
	Location fit-out plan	Ready for occupation '*date*'	Location Team Leader	*High*

Table 4-15 A completed RACI Chart — case study F (note that this is an extract from the RACI Chart, which is several pages in length). R = responsible; A = accountable; C = consulted; I = informed

Project Management Toolkit — RACI Chart

Project: Relocation project | **Date:** Confidential

Names → Activity ↓	Sponsor	Project Director	Project Manager	Functional Team Leaders	Locations Team Leaders	Business Unit Team Leaders
1. Core project delivery plan development	C	A	R	C	C	C
2. Location project delivery plans development	I	C	A	C	R	C
3. Business unit project delivery plans development	I	C	A	C	C	R
4. Functional policy development (finance, human resources, commissioning, facility manager)	—	I	I	R	C	I
5. Overall relocation project time-line (milestone plan)	I	A	R	C	C	C
6. Master move schedule development and management	—	A	C	R Facility Manager	C	C
7. Project cost plan development and management	I	A	C	R Finance	C	C
8. Benefits management	A	C	C Functional benefits	R Location benefits	R Business unit metrics	R

Table 4-16 A Completed Control Specification Table — case study F

Project Management Toolkit — Control Specification Table

Project: Relocation project		**Date:** Confidential
Note that unless stated otherwise, progress reports against each control methodology are to be issued to the Project Manager each month for collation into the one-page high-level progress report and onward circulation as appropriate		
	Cost control	
Cost objective/CSF	**Control methodology**	**Responsibility**
Meets agreed budget (capital and revenue)	Overall project cost plan (based on a master financial model)	Functional Sub-Project Team Leader (financial)
Commissioning cost meets budget	Commissioning contract management	Project Manager
Move cost meets budget	Move company management	Functional Sub-Project Team Leader (facilities management)
Meets agreed budget (capital and revenue)	Location cost plan	Location Sub-Project Team Leader
	Schedule control	
Schedule objective/CSF	**Control methodology**	**Accountability**
New build completion by '*date*' Completion of all moves by '*date*' Site disposals by '*date*'	Milestone plan for overall project schedule Schedule change control process linked to location and business unit teams	Project Manager
Completion of all moves by '*date*'	Logic linked master move schedule for the overall project moves	Functional Sub-Project Team Leader (facilities management)
Completion of all moves (in and out) by '*date*'	Logic linked move schedule within each location sub-team	Location Sub-Project Team Leader
	Scope control	
Scope objective/CSF	**Control methodology**	**Accountability**
Total number of business units to move Total number of people to move within a business unit	Database with change control in use and monthly reports to Project Director	Project Manager
Functional policies	Deliverables listing by function (for example facilities management, human resources, finance)	Functional Team Leaders

Handy hints

Never be afraid to plan

Project Managers are usually focused on delivery and project stakeholders expect this of them — however to miss out the planning stage is the start of being 'out of control'.

Don't progress your project without 'real' sponsor 'buy-in' to your plan

Many projects have failed to deliver what was expected of them because of this lack of organizational support. If your sponsor agrees with your plan he is confirming that it aligns with the business benefits, which he is accountable for delivering. A project delivery plan is therefore a 'contract' between you and your sponsor.

Don't get lost in the tools and techniques of control

Whilst it is useful that there is a mass of software products out there to support the development and management of scope, cost and schedule, it is vitally important that any Project Manager makes a pragmatic decision about the appropriate tools for each specific project. The control toolkit is unlikely to be exactly the same for any two projects — because no two projects have ever been exactly the same.

Planning is so much more than a bar chart

If someone is asked if they have a plan for their project and answers with a reply of 'yes', it usually means they have a (very nicely coloured in) bar chart. Nothing is more disappointing than seeing a Project Manager believing that planning a timeline is sufficient planning. This is one tool and on its own it is no match for the uncertainties which lie ahead in every project.

Always look to expand your Stage Two toolkit

The Stage Two toolkit contains five tools out of many that are actually used in the planning stages of a project. There are other tools that will help you as you define and refine your understanding of what the project plan needs to articulate before delivery commences — search them out and use them appropriately and pragmatically.

Further reading

Project planning is a core project management activity. These aspects are covered in more detail by *Real Project Planning: Developing a Project Delivery Strategy* a further book in this Project Management Series (ISBN: 978-0-75068472-9).

Additionally the websites referenced on page xi can be used to source further 'best practice' material.

And finally . . .

- Ask 'how?' — if you don't know how you're going to deliver the project then how are you going to deliver it!
- Understand the value of a robust plan and also when planning is starting to get in the way of delivery. Be pragmatic!
- Be flexible — use and adapt the tools to achieve your goals.

5 Stage Three: in control?

The third value-added stage in a project is all about the delivery. A project, which is required by the organization, with a robust business case and having a robust delivery plan can still fail if the delivery stage is not controlled. The value in a plan is actually in using it!

During the delivery phase the Project Manager is continually trying to forecast the outcome with increased certainty. These forecasts focus on the three project variables (cost, scope and time) and increasingly on a fourth (business benefit).

In control?

For a Project Manager, being 'in control' is their *raison d'être*. They are not always involved in assessing if a project is required, they may even be handed the project plan *but* they are always there to *deliver!* and to control that delivery.

A typical model of control is shown in Figure 5-1. The essence of control is that by monitoring a situation versus a set of goals, suitable adjustments can be made in order to achieve the desired outcome.

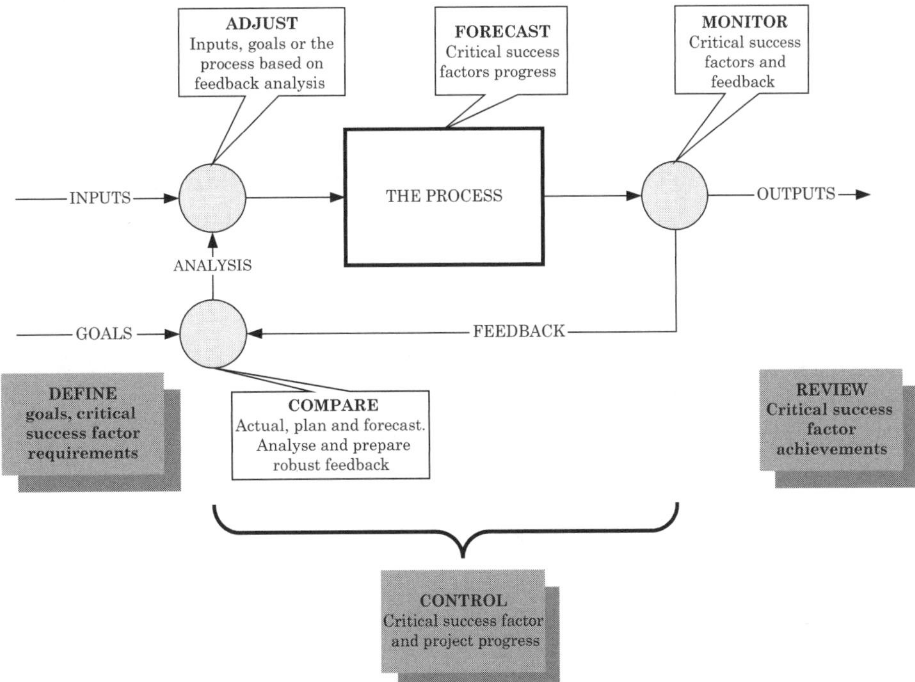

Figure 5-1 A control model

The basic concepts ensuring a project is in control are:

- Planning for a successful outcome — a project should have a robust project delivery plan as outlined in Chapter 4.
- Following the plan — activities should be progressed in line with the project delivery plan.
- Monitoring progress — a control strategy should be used to monitor how the project is progressing versus the plan.
- Forecasting — forecast the overall project outcome based on the progress data.
- Feedback — based on the analysis of the current status versus plan and the impact this has had on the forecast, project activities can be 'adjusted' appropriately in order to achieve project success.

Without a feedback loop any project is only being 'monitored' and not actively managed.

Control strategy

There are many parameters to 'being in control' in a project although most can be categorized into cost, schedule or scope control. For each critical success factor (CSF) you need to be able to operate the control model whilst at the same time operating an overall project control model.

When setting up a control strategy, Project Managers use control hierarchies to ensure alignment throughout their projects (Figure 5-2). In this way they can develop their control strategy and ensure consistent and aligned progress monitoring, outcome forecasting and change control.

The following tools are just a few of those that can be used during the delivery stages of a project to provide support in asking 'are we in control?':

- 'In Control?' Checklist.
- Risk Table and Matrix.
- Earned Value Tool.
- Project Scorecard.

Projects, which do not receive an appropriate level of control during their life, are more likely to fail:

- The level of uncertainty of the outcome will increase — you won't know what's happening!
- The Project Manager and Project Team will be 'fire-fighting' — they will be out of control, reacting to each crisis as it occurs.

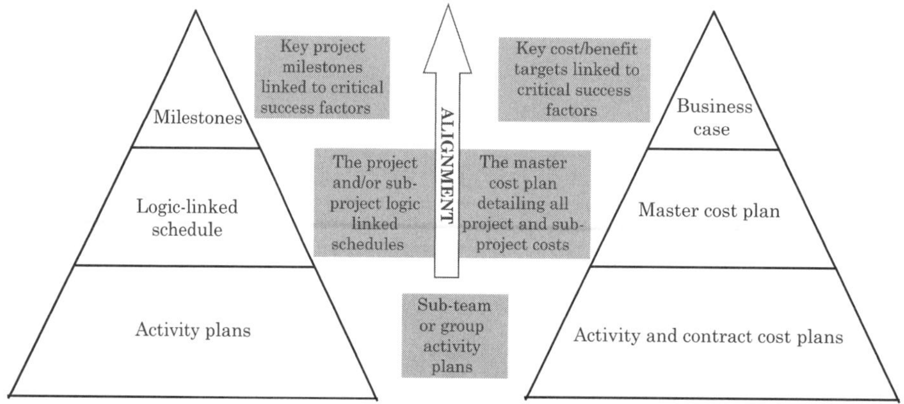

Figure 5-2 Control hierarchies

Tool: 'In Control?' Checklist

If any project (at whatever stage) cannot be challenged by the questions contained in this checklist *and* have robust answers, suitably backed up, then the project is not in control.

This tool can be used at any point in Stage Three as an ongoing 'health-check'.

The 'In Control?' Checklist explained

The checklist contains nine major checks:

- Stage Two check.
- Scope definition.
- Business change management.
- Risk and issue management.
- Project organization.
- Contract and supplier management.
- Project controls strategy.
- Project review strategy.
- Stage Three decision.

Table 5-1 shows the 'In Control?' Checklist with high-level guidance on how to complete each check. The following are additional, more detailed, notes to support checklist completion.

Stage Two check

This check is included so that any changes can be referenced; the following question needs to be asked:

- Have there been any changes since Stage Two completion?

During project delivery the benefits management aspects are firmly within the sponsors responsibilities. So it is important for the sponsor to confirm the business case is unchanged at the start of each Stage Three review.

Additionally, if there have been other Stage Three reviews, it is important that any actions from such reviews are known about so that they can be followed up.

Business change management

These checks establish that the business is keeping its part of the bargain — that it is implementing any changes that the project requires for overall delivery of the project and/or the benefits. The following questions need to be asked:

- What is the current status of stakeholder management?
- How will the business be expected to operate as a result of the completion of the project?
- Is the business ready for the project?
- What is the strategy for handover of the project to the business?

Table 5-1 The 'In Control?' Checklist explained

Project Management Toolkit — 'In Control?' Checklist

Project: *<insert project title>*	**Project Manager:** *<insert name>*
Date: *<insert date>*	**Page:** 1 of 2

Stage Two check

Have there been any changes since Stage Two completion?
<insert any changes that may impact the delivery of the project or the associated benefits since either Stage Two completion or since a previous Stage Three project review>

Business change management

What is the current status of stakeholder management? (Review the original stakeholder map and discuss)
<insert comments on the status of the stakeholder management plan>
How will the business be expected to operate as a result of the completion of the project?
<comment on the main changes to the business as a result of the project>
Is the business ready for the project?
<insert comments on the development and/or delivery of the plans for the management of the business change associated with the project including any sustainability plans>
What is the strategy for handover of the project to the business?
<insert comments on the development and/or delivery of the plans for the handover of the project>

Scope definition

Has the scope changed since Stage Two completion?
<insert details of any changes to the scope since the project delivery plan was approved or a previous Stage Three project review was conducted>
What is the project progress against the Stage Two defined and prioritized project objectives?
<insert a current list of project objectives/CSFs with a comment on their progress versus plan>

Project Roadmap

What project stages/stage gates have been completed? (For example project approved, design complete)
<comment on the internal or external stage gates which have been successfully achieved since Stage Two completion or the previous Stage Three project review>

Risk and issue management

Have all project risks been reviewed regularly during project delivery?
<comment on any high priority risks and the frequency of review during the project>
What is the status of mitigation plans?
<attach a copy of the high priority mitigation plans and their current progress>
What is the status of contingency plans?
<attach a copy of the high priority contingency plans and comment on which have been implemented>
What is the overall likelihood of achieving the project CSFs?
<comment on the chances of achieving the critical path of success>

(continued)

Table 5-1 (Continued)

Project Management Toolkit — 'In Control?' Checklist

Project: *<insert project title>*	**Project Manager:** *<insert name>*
Date: *<insert date>*	**Page:** 2 of 2

Project organization

Are project activities being completed by the appropriate members of the organization?
<insert comment on the Project Team performance in terms of efficiency, effectiveness, capability and capacity and versus the defined roles and responsibilities>

Contract and supplier management

What external suppliers are being used?
<comment on the status of the contract plan>
What is the external supplier status and performance?
<confirm that procedures to manage supplier selection and performance are being followed, comment on supplier performance>

Project controls strategy

Are project costs under control? (Review cost plan — for example actual versus budget)
<comment on the current status of the cost plan, how regularly is this being reviewed, analysed and reported?>
What is the likelihood that the project budget will be maintained (forecast to completion?)
<insert a current forecast of the final project cost versus the authorized budget>
Is the project schedule under control? (For example review schedule and milestone progress)
<comment on the current status of the schedule, how regularly is this being reviewed, analysed and reported>
What is the likelihood that the project schedule will be achieved (forecast to completion?)
<insert a current forecast of the final project completion date versus the authorized schedule>
Are there any changes to scope (quantity, quality and functionality)? Are the costs and schedule under control?
<comment on the status of the change control process, attach a copy of the current change register>

Project review strategy

Are regular Stage Three reviews being conducted? (Is performance being managed and monitored?)
<comment on the schedule of reviews which have been planned and conducted, attach the previous review summary>
Is project performance adequate for project success?
<comment on project performance based on previous project review data plus this review>
Is there regular reporting? (Is the Project Team adequately managing communication of progress and performance to all stakeholders?)
<comment on the adequacy of progress reporting and communication to all internal and external stakeholders>

Stage Three decision

Is the project under control? (Is the project control strategy robust enough for project delivery to continue?)
<comment on the success of the control strategy in use>
What is the certainty that the project will be successful?
<comment on the probability that this project will achieve its goals>

Stakeholder management remains a critical aspect of maintaining control in this area of the project and the plan, plus any associated communication agreements, should be reviewed. A key part of the communication with stakeholders in the business is addressing change — change within the project and also change within the business:

- Changes within the business need to be understood so that the impact on the project can be assessed, for example if specific business changes cannot be completed in time.
- Changes within the project may further impact the business and cause more or less change. The business needs to understand and challenge this as appropriate.

Scope definition

Being in control of the scope is at the core of being in control of a project. Being on-time and on-budget mean nothing if the scope doesn't meet quality, quantity and functional requirements. A defined scope delivering clear business benefit is at the heart of any business case, as well as time and cost considerations. These checks therefore continue to challenge whether the required scope is being delivered:

- Has the scope changed since Stage Two completion?
- What is the project progress against the Stage Two defined and prioritized project objectives?

Project objectives should be tracked; there are many tools available to do this, for example Design Deliverable Listings and Procurement Plans (see *Project Delivery: Managing your Project in Control* as detailed in the further reading section at the end of this chapter) as well as overview tools such as the Table of Critical Success Factors (see page 70), which is easily converted into a tracking report. The reports must track the quality and functional attributes as well as quantity, cost and time. Methodologies such as earned value monitoring can be usefully used to track scope in this way if the system is set up to review all scope parameters, although it is more typically associated with cost and schedule tracking.

The major reason for scope being 'out of control' is usually related to the lack of change control — Figure 5-3 describes a very typical process.

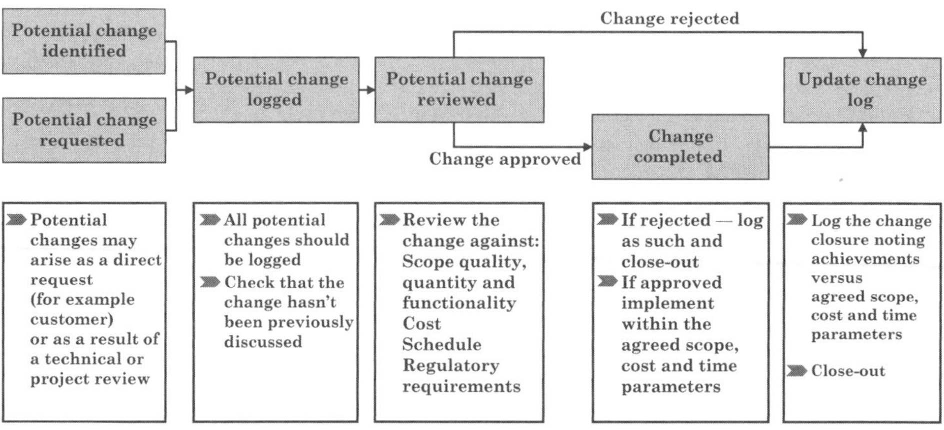

Figure 5-3 A simple change control process

Project roadmap

This check asks the question below and confirms the position in the project roadmap, for example is the design complete? Has everything been procured?

- What project stages/stage gates have been completed?

Without some form of roadmap it can be difficult to explain to external stakeholders 'where you are' and it may also be difficult to keep on top of 'where you are going'.

Risk and issue management

These four questions are at the core of checking whether uncertainty is being managed — where are the uncertainties? Are there plans to deal with them? Ultimately this allows for a review of whether the overall project is likely to achieve success:

- Have all project risks been reviewed regularly during project delivery?
- What is the status of mitigation plans?
- What is the status of contingency plans?
- What is the overall likelihood of achieving the project CSFs?

Risk management is a continuous process throughout any project and this is described in more detail on page 110 with the introduction of a typical risk tool — the Risk Table and Matrix.

Project organization

This check addresses the 'hard' and 'soft' aspects of Project Team behaviour. It also allows a review of the team capability and capacity needs.

- Are project activities being completed by the appropriate member of the organization?

From a 'hard' perspective, to be in control of the team means that they are doing the role which was planned in the way it was planned. For example a senior engineer cannot delegate the final checking of key calculations which assure the safety of a critical system; he is responsible for this activity and was allocated this task due to his chartered engineer status.

From a 'soft' perspective, to be in control of the team means that the team are capable, have the capacity to do their role and they are motivated to do so. Maintaining team motivation is important for any Project Team (Figure 5-4):

- Communication — good internal project communication is a key aspect of maintaining a motivated team. If they are aware of what's going on they can better understand their role in it.
- Feedback — constructive and fair positive and developmental feedback supports effective team and individual motivation. This needs to be balanced so that feedback is taken on board and acted upon.
- Involvement — if teams are involved in generating the solutions, which they must deliver, then they are more likely to buy-in to the solution. As appropriate the teams' knowledge must be used to support the project in this way. This is a key step in maintaining a shared goal.
- Shared goal — at the heart of team motivation is the ability to maintain a shared goal. A team made up of individuals all moving in the same direction is likely to be 'more than the sum of the parts'.

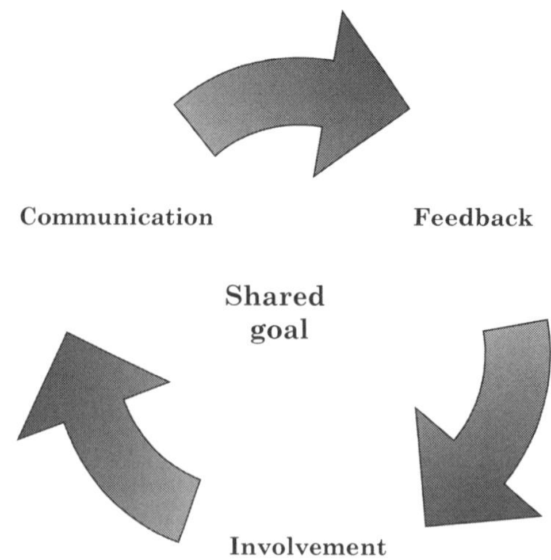

Figure 5-4 Maintaining team motivation

Contract and supplier management

These checks review the contract strategy — who is being used and how will they be managed?

- What external suppliers are being used?
- What is the external supplier status and performance?

A contract plan will detail why external support is needed in that area and then the type of supplier needed and the type of contract proposed. To be in control the contract plan should be used as a 'live' control tool to assess the performance of the plan as the project progresses.

Project controls strategy

These checks confirm the Project Manager's ability to forecast that the project will deliver the scope, cost and time to agreed targets:

- Are project costs under control? (For example review cost plan — actual versus budget)
- What is the likelihood that the project budget will be maintained? (Forecast to completion?)
- Is the project schedule under control? (Review schedule and milestone progress)
- What is the likelihood that project schedule will be achieved? (Forecast to completion?)
- Are there any changes to the scope? (Quantity, quality and functionality) Are the costs and schedule under control?

The type of information to support a cost forecast will typically be some form of cost report (Table 5-2) and in addition a method to show that the contingency fund is being appropriately managed (Figure 5-5).

Table 5-2 is a simple example of a contract cost report that looks at what funds have been committed against specific items in the budget. Based on past level of contract variations, progress through the contract and an assessment of risk related to that contract, a forecast of costs to complete can be made and therefore the total likely cost (Y). Once completed for all items the total project cost can be forecast by adding the contingency fund to this figure.

Table 5-2 A typical contract cost report

Order number	Item	Budget	Order value	Approved variations	Forecast to completion	Total forecast cost	Comparison with budget	Percentage progress
	Contract name	Authorized funds for order (X)	Agreed order price (A)	Agreed contract variations (B)	Project management forecast of additional spend (C)	Y = A + B + C	Y/X as a percentage	Percentage contract completion

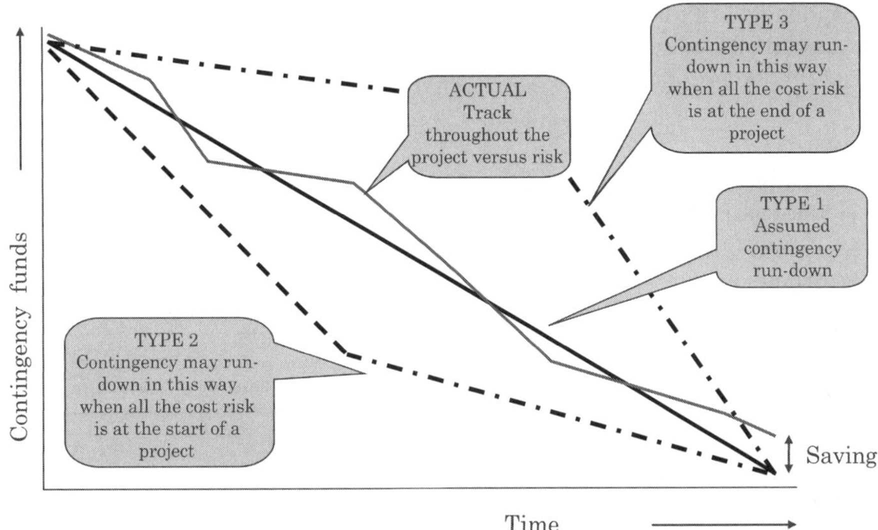

Figure 5-5 Contingency run-down

Contingency should always be maintained as a separate fund even though the total amount needed is calculated based on the risk of every item in the project (cost, schedule and scope risks can all be converted to a cost amount). Contingency can then be tracked as the project progresses versus a planned expenditure which is one of three types:

- Type 1 — in the absence of a contract strategy contingency run-down may be assumed to be linear. In this case the risk is assumed to be equally distributed throughout the project.
- Type 2 — when reimbursable and measured term contracts are used the risk is all at the start of the project, based on choosing the right supplier and negotiating the unit costs.
- Type 3 — when using large fixed price sub-contracts during a construction phase, variations and change agreements are more likely towards the end of the contract.
- Actual — as contingency is either allocated (used versus a specific risk which then occurred) or released (not required as the specific risk has been eliminated) the contingency fund should be tracked.

The type of information to support your schedule forecast will typically be some form of bar chart or Milestone Progress Report (Table 5-3).

More complex reports can be used to monitor schedule, for example earned value and 'S' curves (see page 124).

Project review strategy

These checks confirm if project performance is being managed and is in control:

- Are regular Stage Three reviews being conducted?
- Is there regular reporting?

Reporting performance on a continuous basis throughout the project is important. However the format, content and length of the report must be appropriate for both the audience and also the way that the data in the report is to be used. Monthly reports that take 3 weeks to generate do not perform any useful control function other than to record history.

Table 5-3 A typical Milestone Progress Report

Project: *<project name>*		**Date prepared:** *<this would usually be prepared at the month end in conjunction with the monthly progress report>*		
Report prepared by: *<Project Manager's name>*		**Schedule reference:** *<the document reference, version and issue date for the schedule to which this data is connected>*		
Milestone progress				
Ref	**Activity**	**Planned date**	**Actual or forecast date**	**Comment**
	<specific milestone activity>	*<date>*	*<date>*	*<comment on delay and/or associated impact>*
				<generally this would not be longer than a page>
	<completion date>			*<always report on forecast end date>*

General Comments

The above is an *example* format for a 'Milestone Progress Report'. This is a key document that should be used to identify and log activities that are critical to the success of the project. Progress against these milestones must be monitored. This report also provides a high-level review of the project status and needs to be accurate and up-to-date.

- The activities monitored will vary from month to month *but* every time an activity is put on the list it *remains* so that its progress is monitored. It can be deleted the month after it has been reported as complete.
- The '*Ref*' column should be used to note the activity reference from the associated high-level schedule (usually referred to as the activity code).
- The '*Planned*' dates should be the original date and therefore *never* change from the original schedule.
- The '*Actual /forecast*' date should be either the date when the activity was 100% completed or the current forecast of the date when it will be completed.
- The '*Comments*' column should be used to record *change order references* and/or comments if the forecast date is different from the actual date. In any event the impact of changes should be assessed and reported.

Stage Three decision

Once all available project delivery status data has been gathered and analysed the following checks need to be addressed:

- Is the project under control?
- What is the certainty that the project will be successful?

The end point for each Stage Three review is confirmation that the project delivery is in control, thus reaffirming that the project control strategy is robust enough for project delivery to continue. Any project reviewer should equally be capable of confirming if, in their opinion (backed up by the project data), the project is likely to be successful. Without this final assertion what is the point of the review?

Using the 'In Control?' Checklist

This tool can be used in many different ways:

- To formally or informally review the project progress within the Project Team.
- To support a formal independent external audit of the project.
- To support an informal external view of the project.

The ultimate aim of any type of review is to answer one question:

- Is the project in control? (Stage Three decision)

Type of project review		Who conducts the project review: INTERNAL	Who conducts the project review: EXTERNAL
	FORMAL	**PROJECT TEAM MEETINGS** There should be formal meetings at all levels in the project organization which address progress, performance, issues and risks These should be based on accurate project data and robust analysis	**PROJECT HEALTHCHECK** This is an independent project review A project should be available for this type of review at any point in its life-cycle
	INFORMAL	**NORMAL PROJECT TEAM BEHAVIOUR** The Project Manager and the project team are constantly reviewing the potential for the project to meet its objectives A culture of informal discussions with the team and within the team should be generated	**SPONSOR PROGRESS REVIEW** Sponsor project discussions should take place regularly The Project Manager should also have a senior project management mentor who can give an independent view on the project performance

Figure 5-6 Project review matrix

Figure 5-6 shows the various types of review through the use of a two-by-two matrix. Each review is made up of a different team both internally and externally.

To complete the checklist the review team need to have access to the data that supports the demonstration of a project in control:

- 'Hard' data related to the control strategy (scope, cost, time).
- 'Soft' data such as stakeholder feedback and team performance.

Each of the case studies in this chapter uses the 'In Control?' Checklist to demonstrate the various uses of the tool. In completing the checklist it can be seen how the use of other tools within this chapter can support the demonstration of a project in control and therefore the completion of the Stage Three decision.

Tool: Risk Table and Matrix

The aim of this tool is to support the effective delivery of a project by identifying and managing uncertainty; risks to the success of the project.

Project Managers have always tried to identify the risks at the start of a project — as a part of project delivery planning. This tool is equally applicable to the development of a robust delivery plan as outlined in Chapter 4. However, the ongoing value of risk assessment is in actually using risk management as a core part of the control strategy to analyse:

- Cost risk.
- Schedule risk.
- Scope risk.

Risks change and the aim of this tool is to provide a mechanism to track those changes and to make appropriate risk responses. The Risk Table and Matrix is a control tool which is why it has been placed in Chapter 5; however a baseline risk assessment clearly has to be set up in Stage Two prior to project delivery kick-off.

The Risk Table and Matrix explained

The tool contains three main sections:

- Risk description.
- Risk assessment.
- Action planning.

Table 5-4 shows the Risk Table with high-level guidance on how to complete each section. The following are additional notes to support tool completion.

Table 5-4 The Risk Table explained

***Project Management Toolkit* — Risk Table**						
Project: *<insert project title>*				**Date:** *<insert date>*		
Risk description			**Risk assessment**		**Action planning**	
Risk number	**Risk description**	**Risk consequence**	**Occur?**	**Impact?**	**Mitigation plan**	**Contingency plan**
Number all risks for use in Risk Matrix	Describe what could happen	Describe the consequence of the risk	Assess if low, medium or high probability of risk occurring	Assess if low, medium or high impact on project goal	Describe the actions which would either stop the risk occurring or minimize the impact	Describe the contingency plan which would minimize impact if the risk occurred
<insert risk number>	*<describe the risk scenario>*	*<describe the impact — what happens?>*	*<insert risk probability>*	*<insert risk impact>*	*<describe the mitigation action plan>*	*<describe the contingency action plan>*

Risk description

The aim of this section is to identify all possible risk scenarios — how the scenario would occur and then what the impact would be on the project.

If a high-level critical path of success has been developed then the risks should be categorized against each of the Level 1 CSFs (see page 70).

Risk assessment

In order to assess each risk scenario a scoring system needs to be developed. The scoring system should only be as precise as the risk scenario data being used — there may be a low chance of a risk occurring so to actually assign this a percentage probability may be inappropriate:

- Scoring risk probability — usually a low, medium, or high scoring is appropriate for most risk scenarios due to the accuracy of the data used to generate the scenario. Low means less than 50%; medium is a 50% probability and high is greater than 50%.
- Scoring risk impact — it is usual to relate this to either the specific CSF category being reviewed or the impact on the project as a whole — low means that it does not significantly impact project success; medium means it will have an impact and high means that it will prevent project success.

Once scoring is completed the risks should be plotted on the Risk Matrix (Figure 5-7).

Action planning

Figure 5-7 uses a 'traffic-light' system for risk response actions:

- Green — a low priority risk. Mitigation plans should be planned but not implemented.
- Amber — a medium priority risk. Mitigation plans should be developed and only implemented if sufficient resources (not involved in implementing red risks) are available.

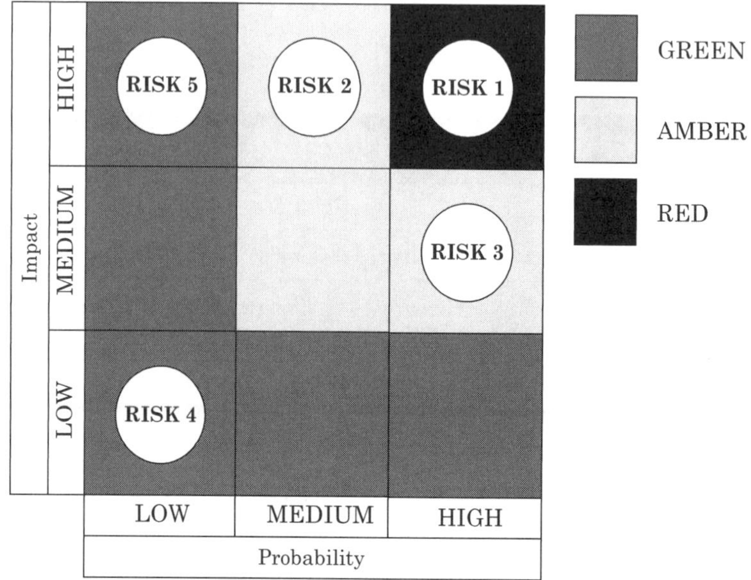

Figure 5-7 The Risk Matrix

- Red — a high priority risk. Mitigation plans are a high priority and should be robustly developed and implemented as soon as possible. Contingency plans should be designed and be ready for implementation should the mitigation plans prove to be unsuccessful.

The Risk Matrix supports the view that it is not pragmatic, nor necessary, to eliminate all possible project risks. No project has a limitless resource pool in terms of funds or people.

The action planning section of the Risk Table should be used to highlight the 'traffic-light' colour of a risk so that the table itself represents a simple visual overview of the status of the project risks. Additionally, the decision regarding which plans are to be progressed should be inserted along with a summary of the plan.

Using the Risk Table and Matrix

A risk is a potential issue that could significantly impact the achievement of the project objectives if it occurred. Once a risk is identified, it needs to be assessed:

- What is the likelihood that this risk will actually occur — low, medium or high?
- What is the potential impact if the risk did occur — low, medium or high?

Then two types of plan need to be developed:

- Mitigation plan — an action plan to eliminate the possibility of the risk occurring.
- Contingency plan — a back-up plan to use when the risk has occurred which minimizes the impact on the project.

The process of risk assessment and continuous management throughout all project phases is described in Figure 5-8.

A risk management process is best used in a team situation. Initially, for the first risk assessment at the project delivery planning stage, the team should brainstorm the types of risks which could prevent project success. If a project has already identified a critical path of success (see page 70) then this has effectively identified the major risk categories, for example risks that prevent achievement of CSF Level 1 (1), risks that prevent achievement of CSF Level 1 (2) and so forth. In the risk session the team should:

- Brainstorm the risk scenarios for each risk category. Once this is completed the risk scenarios should be transferred to a Risk Table.
- Agree a scoring basis — this is important as one person's perception of a high risk may not be the same as another's.
- The risk scenarios should then be reviewed in turn and the impact and probability score agreed. Discussion here is important to ensure that the risk scenario and consequence has been adequately defined.
- The results should be plotted on to a Risk Table. Mitigation plans and contingency plans should then be developed for all red risks. Depending on team resources mitigation plans can be proposed for amber risks.

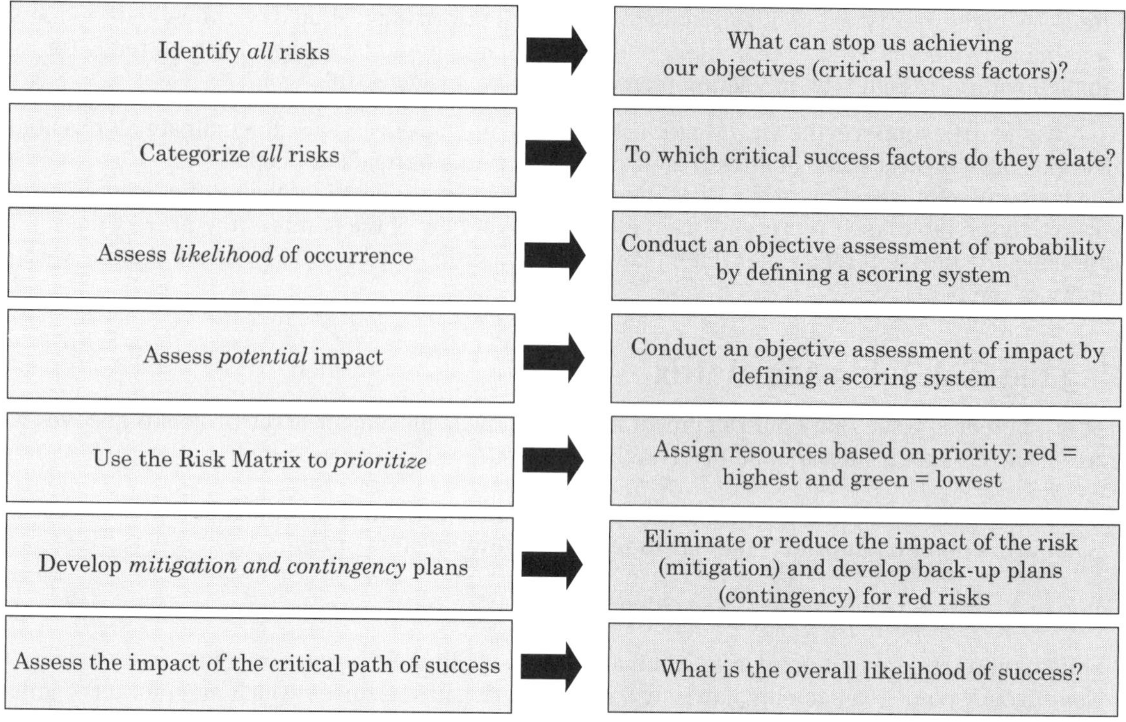

Figure 5-8 A risk management process

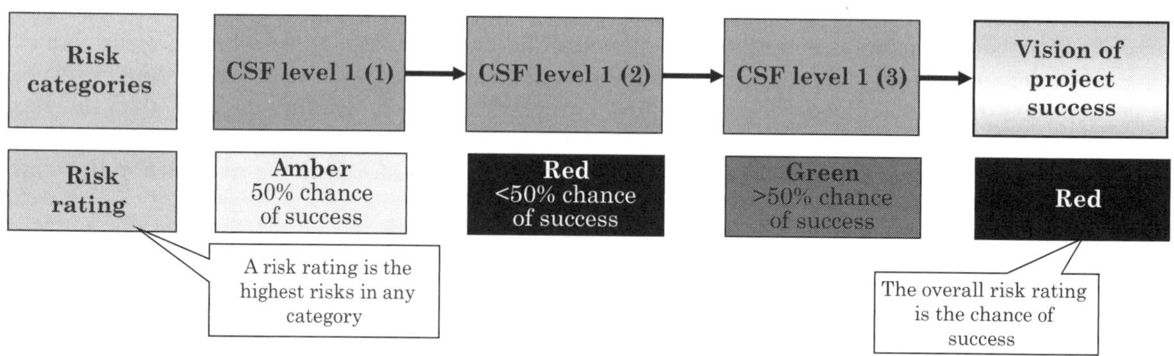

Figure 5-9 The critical path of success and risk

At the end of the risk session it is always useful to review how many red, amber and green risks there are against each CSF (risk category) and to assess what this means for the overall success of the project. In this way a 'risk rating' for a CSF and for the overall project can be determined (Figure 5-9), for example if CSF Level 1 (1) has three risk scenarios, two amber and one green then the overall risk rating is amber. The risk rating is equal to the highest risk scenario rating for that CSF. The same is true of the project risk rating — it is equal to the highest risk rating of any CSF. Remember the vision of project success can only be achieved if all the CSFs are achieved (see page 70).

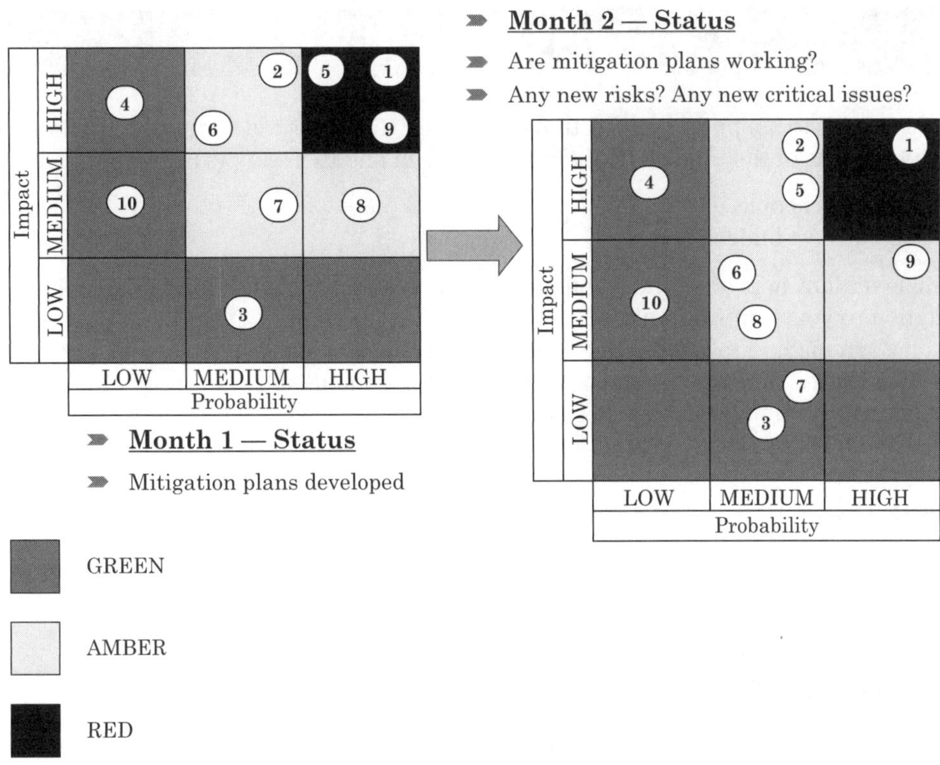

Figure 5-10 Managing risk

Risk management can then become more structured and systematic for the team and it is usual for the team to continue to meet for risk reviews at a regular frequency throughout the delivery of the project.

At a high-level the team can review the Risk Matrix which gives a good visual representation of risk movement (Figure 5-10):

- Mitigation plans success should be obvious — for example have the risks moved from red to amber to green? Do contingency plans need to be kicked-off?
- The need to develop new contingency or mitigation plans can be assessed from the movement of risks from green to amber to red.

The team should also assess if additional risks are likely and if risks have actually occurred and become 'live' project issues. The team should also assess the ongoing impact of the risk situation on the probability that the project will be successful (Figure 5-9).

Tool: Earned Value

The aim of this tool is to support identification of whether the project is in control. It allows a detailed review of critical areas of the project, or the whole project, and then supports an assessment of:

- Progress versus schedule.
- Progress versus cost budget.

Earned value initially appears to be a complex concept. However, if used pragmatically, it can be a valuable addition to your control toolkit.

Earned value is a method to assess the progress of a project against tangible deliverables rather than just reviewing total time or cost spent. To support the comparison of earned value to cost or time spent, conversion into common units is necessary.

The Earned Value Tool explained

The tool contains three main sections:

- Earned value basis.
- Earned value data collection.
- Earned value interpretation.

Table 5-5 shows the Earned Value Tool with high-level guidance on how to complete each section. The following are additional notes to support tool completion.

Earned value basis

Earned value is a method to track the progress of completed work without linking it to actual costs. It is therefore important that a robust method of assessing progress can be made which is independent of man hours or costs (as the two are both methods of assessing actual costs), for example number of design deliverables, length of pipe-work installed, or volume of concrete laid.

As the progress data will need to be compared to either cost, man-hours or percentage for planned and actual the progress data needs to be converted into the same units. This is the earned value calculation.

Earned value data collection

A plan versus time should be developed (which links to the overall project plan) and then at regular intervals the actual cost and the earned value cost should be collected.

Earned value interpretation

The classic way to interpret earned value is to construct an 'S' curve (Figure 5-11). This is a curve that plots cumulative costs (or other progress measures) over time. It therefore represents a link between the project schedule and the project costs.

Table 5-5 The Earned Value Tool explained

***Project Management Toolkit* — Earned Value Tool**			
Project: *<insert project title>*		**Date:** *<insert date>*	
Earned value basis			
<insert a description of the project area which is to be tracked by earned value, this could be the whole project, a sub-project, a CSF or an activity within a CSF>			
Cost management basis	*<insert how actual costs are to be measured for this project area>*		
Progress management basis	*<insert how actual progress is to be measured for this project area — for example the tangible deliverables>*		
Earned value calculation	*<insert the method to be used to calculate the earned value in the project area — for example how the tangible deliverables will be converted to an equivalent cost>*		
Earned value data collection			
Timeline	**Plan**	**Actual**	**Earned value**
<insert date>	*<insert planned cost spend by this date>*	*<insert actual cost spend by this date>*	*<insert earned value by this date>*
<insert date>	*<insert cost>*	*<insert cost>*	*<insert earned value>*
Earned value interpretation			
<attach a copy of the 'S' curve from the above data and comment on the status of this CSF versus cost and schedule target>			

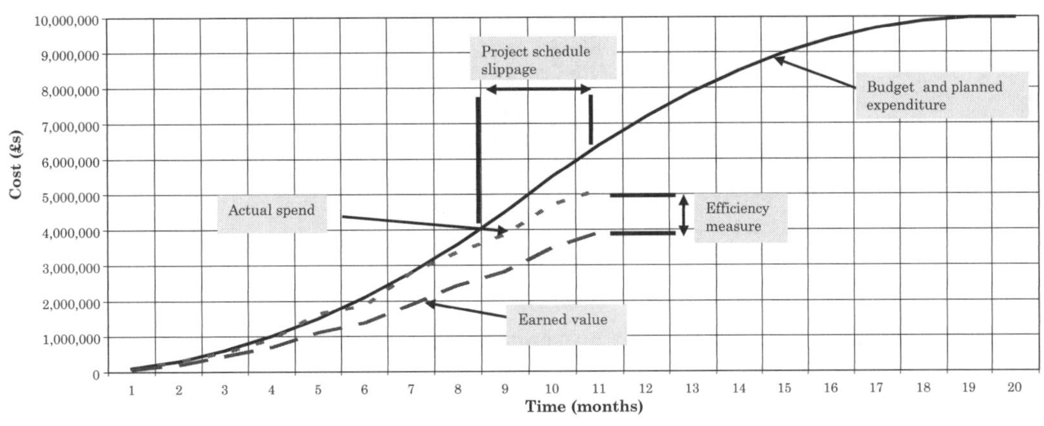

Figure 5-11 The earned value 'S' curve

Three 'S' curves can be plotted: planned, actual and earned value:

- The difference between the actual and the earned value is a measure of efficiency — this can be used to forecast final costs.
- The difference between the planned and the earned value is a measure of schedule adherence — this can be used to forecast the schedule outcome.

Using the Earned Value Tool

There are many available texts that explain how to develop earned value systems, some more complex than others. However, the Project Manager and his senior team must be able to understand, manage and interpret the system used — it is no good having a control engineer develop a nice graph if no one understands it; that it not the basis for a robust control strategy.

It is usual for the Project Manager to develop his own 'S' curve, which will give him the ability to forecast cost and schedule variances. His team may develop their own 'S' curves in order to provide the Project Manager with crucial progress information on critical areas of the project.

The example in Table 5-6 and Figure 5-12 is taken from a design sub-project, which was cycling out of control. By the time the review was completed at the end of July the sub-project team had no way to recover the situation and both cost and time were lost. The tool was used to historically review what happened and showed that the designer was never as efficient as planned right from the start and that by May the programme had already started to slip.

The revised forecast end-date was October with an additional £30 000 agreed. The contract was converted to a fixed price with a bonus/penalty clause associated with the newly agreed end-date.

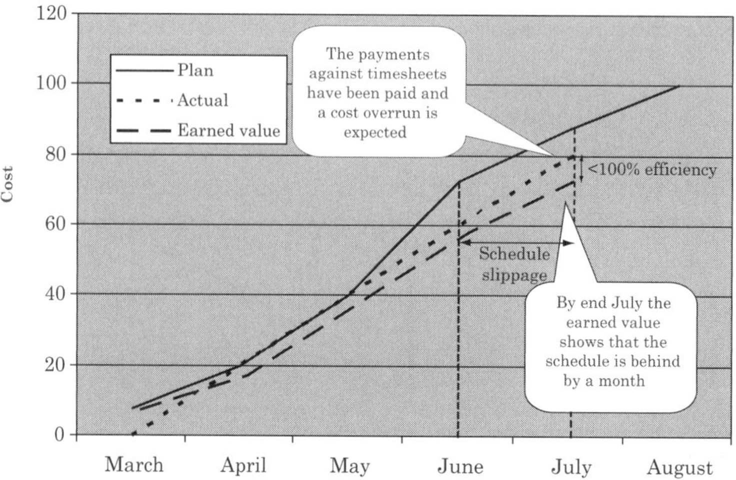

Figure 5-12 Earned value 'S' curve example

Table 5-6 The Earned Value Tool — an example

Project Management Toolkit — Earned Value Tool

Project: New office complex	**Date:** July

Earned value basis

The design of the office complex is critical to the success of this project and is to be tracked closely to ensure that the schedule and costs are maintained as well as design quality

Cost management basis	The design is sub-contracted to an external designer on a target-reimbursable basis (£100 000). 200 deliverables are required to a pre-agreed standard within 6 months of the project's start. Monthly interim payments against approved timesheets have been agreed
Progress management basis	Only 100% completed deliverables are to be included within the progress measurement. 100% includes for draft issue, comments and subsequent revision prior to approval
Earned value calculation	The total number of deliverables used to confirm progress will be converted to an equivalent cost by comparison between total deliverables (200) and total cost (£100 000) — £0.5000 is earned per complete deliverable

Earned value data collection (all figures are cumulative)

Timeline	Plan (planned cost spend)	Actual (actual cost spend)	Earned value
March	£7.5000 = 15 deliverables	0	10 deliverables = £5 000
April	£20 000 = 40 deliverables	£20 000	30 deliverables = £15 000
May	£40 000 = 80 deliverables	£40 000	70 deliverables = £35 000
June	£72.5000 = 145 deliverables	£60 000	112 deliverables = £56 000
July	£87.5000 = 175 deliverables	£80 000	140 deliverables = £70 000
August	£100 000 = 200 deliverables		

Earned value interpretation

The plan required 175 deliverables to be completed by this July review. Although it was quickly apparent that a delay had occurred (only 140 deliverables had actually been completed) a complete overview required the earned value analysis. See Figure 5-12. This confirms previous progress reviews — the schedule continues to slip due to a combination of inefficient working and too much interaction with the end user which is slowing down final deliverable completion. The design team also stayed the same size even though the workload during June was expected to increase to meet the agreed end date

Tool: Project Scorecard

The aim of this tool is to support the succinct collation of summary project data. In this way it can support:

- Communications — through the development of aligned, clear key messages.
- Project control strategy — through accurate reporting of the current situation.

Reporting is a key part of the above two project activities and although projects will inevitably generate a lot of progress data it is the Project Manager who has to distil that data and convert it into useful information that answers:

- How is the project doing?

The Project Scorecard explained

The tool contains two main sections:

- Project status summary.
- CSF tracking.

Table 5-7 shows the Project Scorecard with high-level guidance on how to complete each section. The following are additional notes to support tool completion.

Project status summary

This will probably take up 50% of the one-page report and it is up to the Project Manager to decipher the appropriate high-level messages for each of the six individual sections. It is important that the Project Manager forecasts on the likelihood of success versus the plan.

Bear in mind that anybody reading this report will have the previous report so that any 'key activities in the next period' which appeared in the last should appear as 'key activities completed' in this report. In this way credibility in a summary report can be achieved.

CSF tracking

The remaining 50% of the one-page report is for 'hard' data. On the basis that the project cannot be successful without achieving the CSFs, this section is for the progress tracking of those CSFs. A traffic-light system is proposed, although the exact meaning of each 'light' should be defined beforehand, for example amber progress could represent a 2-week delay. The risk rating of each CSF should also be tracked and any summary comments against either progress or risk made.

Using the Project Scorecard

The Project Scorecard is essentially a Project Manager's tool. However, it can also be cascaded to sub-Project Team leaders within a project and a hierarchy of connected scorecards are easily developed thus ensuring alignment of progress information. The scorecard can be used to support:

- Internal Project Team communication — so that all parts of the team are aware of the overall status of the project.
- External stakeholder communication — so that stakeholders are kept informed and involved.

Table 5-7 The Project Scorecard explained

<table>
<tr><th colspan="5">Project Management Toolkit — Project Scorecard</th></tr>
<tr><td colspan="2">Project: <insert project title></td><td colspan="3">Date: <insert month and year></td></tr>
<tr><th colspan="5">Project status summary</th></tr>
<tr><td>Project status</td><td colspan="4"><insert summary comments on the project status, for example current stage in the project roadmap, stakeholder or team summary></td></tr>
<tr><td>Project control status</td><td colspan="4"><insert summary comments on the overall risk profile, overall cost and schedule forecasts and any plans which need to be put in place to regain or maintain control></td></tr>
<tr><th colspan="3">Key activities in this period</th><th colspan="2">Achievements</th></tr>
<tr><td colspan="3"><insert the most important activities completed this month — activities started or finished, stage gates passed></td><td colspan="2"><insert the most significant milestone achievements or CSF tracking></td></tr>
<tr><th colspan="3">Key activities in next period</th><th colspan="2">Critical issues</th></tr>
<tr><td colspan="3"><insert the most important activities to be completed next month — what are the critical milestones which need to be tracked to keep driving the project forwards></td><td colspan="2"><insert risks which are becoming potential issues to the overall project success, for example funds, resources, skills, people, time></td></tr>
<tr><th colspan="5">CSF tracking</th></tr>
<tr><th>CSF Level 1</th><th colspan="2">Progress</th><th colspan="2">Risk rating</th></tr>
<tr><td>CSF a</td><td>Green</td><td><insert comment></td><td>Amber</td><td><insert comment></td></tr>
<tr><td>CSF b</td><td>Green</td><td><insert comment></td><td>Red</td><td><insert comment></td></tr>
<tr><td>CSF c</td><td>Amber</td><td><insert comment></td><td>Amber</td><td><insert comment></td></tr>
<tr><td>Overall project</td><td>Amber</td><td><insert comment on overall progress status></td><td>Red</td><td><insert comment on overall risk status></td></tr>
</table>

The type of data and key messages that need to be communicated externally may be slightly different from internally. It is up to the Project Manager to consider what is appropriate for the audience.

It is also worth noting that this summary one-page report is not the only report a Project Manager will produce. In order to be 'in control', and to demonstrate that control, many types of progress reports will be used based on 'hard' data:

- Cost reports.
- Milestone schedule reports.
- Project and sub-project 'S' curves.
- Contract summary reports.

Any Project Manager needs as much or as little data as they require to interpret progress, forecast the project outcome and to communicate key messages.

Case study G — if only the 'In Control?' Checklist had been around then!

This case study is based on a real example of a project, which, although eventually completed, did not meet its original time and cost objectives. The 'In Control?' Checklist is used here to trouble shoot — 'why was this so out of control?'

Production plant upgrade

Situation

The project required a large-scale upgrade and automation of an existing manufacturing facility containing four individual process units. Front-end design had been completed in-house and an engineering design contractor selected for the project design on the basis of a target maximum fee. The project delivery strategy was for detailed process and engineering design to be completed by the engineering design contractor with on site resources completing installation. Overall project management was to be completed by a site-based team employed directly by the manufacturing company.

On appointment of the contractor a meeting of all key stakeholders was held at the engineering contractor's office to draw up a project schedule. This included the site manufacturing group who quickly pointed out that all four process units could not be shut down simultaneously and that one could not be shut down at all for more than a few days because of sales requirements which did not allow for stockpiling of material to cover a period of shutdown. They also requested that a single element of the project be delivered very early on in order that it be available for manufacture of a new product.

Initial thoughts were that, as the overall scope of the project had not changed there had not been a significant change to the project. The impact of the manufacturing request was not reviewed and the engineering contractor continued with development of the project.

The project

The project was split up into the four elements by product. The first of these was then split into three phases in order to remove the need for a plant shutdown. The item requested for early delivery was progressed at speed by the engineering contractor. No change control process was followed and the impact of the additional phasing requirements of the project were not considered, even though the whole focus of the project had shifted from one single project to more than six phased much smaller ones.

The site engineering group had resourced the project based on completion of the works as a single element and these resources quickly became overstretched as they attempted to deliver early site construction works as well as continue with design of other elements. This continued due to the phasing requirements of the project which required the site team to be involved in both design and installation elements of the project at the same time. This problem was exacerbated by the fact that a lot of the works required for initial phases were some of the large infrastructure elements of the project required by all elements. The site's engineering involvement in overall project management activities for the design elements were extremely limited. The engineering contractor continued with design works with little or no site project management input. Their only link was provided from the site process engineering team. Change control discipline was not maintained and many changes took place on a verbal instruction with cost implications not understood until the change was complete.

The outcome

The project design costs increased significantly due to both the phasing requirements and change. The site installation for two of the products was not completed, as by the time the design works were completed the products were no longer required. Later investigation identified that their business case had always been questionable. The timescale for the project also overran by 100% not allowing for the additional time that would have been required had the missing construction elements been completed. A formal review of the project was completed in order to learn from the experiences (Table 5-8).

Table 5-8 A completed 'In Control?' Checklist — case study G

Project Management Toolkit — 'In Control?' Checklist

Project: Production plant upgrade	**Project Manager:** Confidential
Date: Confidential	**Page:** 1 of 3

Stage Two check

Have there been any changes since Stage Two completion?
This checklist has been completed as part of a formal review exercise following close-out of the project to assist in the assessment of what happened within the project

Business change management

What is the current status of stakeholder management? (Review the original stakeholder map and discuss)
Stakeholders were not identified early enough within the project and therefore their input was not sought in the early project stages. Stakeholders were not involved until the engineering contractor was in place. The management of stakeholder communication and input was planned and agreed. However this plan was not properly implemented by the site project management team as their resources were overstretched due to early start of construction activities on site
How will the business be expected to operate as a result of the completion of the project?
The operation of the companies main manufacturing asset will be changed by the automation element of the project. This will reduce operator head count and also reduce the time operators spend with the physical plant as operations are controlled from a central control room
Is the business ready for the project?
As has been shown by the non-completion of the upgrade of two of the production processes, the business could only support approximately half the project to completion. The design work for these non-constructed elements was completed which was wasted effort with a significant cost tag attached
What is the strategy for handover of the project to the business?
There is a large training requirement, which is to be completed by a separate site technical support group who will also commission the plant. The project was handed over to this group, by the site project management team, following physical completion of the plant and equipment

Scope definition

Has the scope changed since Stage Two completion?
Although the change was not appreciated at the time, the project changed significantly due to the phasing requirements. This resulted in additional costs for all elements of the project and overstretching of the site project management resource, which was insufficient for the revised project
What is the project progress against the Stage Two defined and prioritized project objectives?
Not applicable — the project is now complete
Although the project was under budget at the time it was brought to close, half of the installation works were not completed. Design elements of the project were approximately 25% over budget

Project Roadmap

What project stages/stage gates have been completed? (For example project approved, design complete)
Project is complete

(continued)

Table 5-8 (Continued)

<table>
<tr><th colspan="2">Project Management Toolkit — 'In Control?' Checklist</th></tr>
<tr><td>Project: Production plant upgrade</td><td>Project Manager: Confidential</td></tr>
<tr><td>Date: Confidential</td><td>Page: 2 of 3</td></tr>
<tr><td colspan="2">Risk and issue management</td></tr>
<tr><td colspan="2">Have all project risks been reviewed regularly during project delivery?
Any risk review or mitigation plan was not completed during the project because of the overstretching of the site project resources who were 'fire-fighting' on the project almost from day 1
What is the status of mitigation plans?
Not applicable
What is the status of contingency plans?
Not applicable
What is the overall likelihood of achieving the project CSFs?
Not applicable</td></tr>
<tr><td colspan="2">Project organization</td></tr>
<tr><td colspan="2">Are project activities being completed by the appropriate members of the organization?
Insufficient project management internally and externally. Everybody was stretched causing de-motivation of the Project Team. For example engineers were completing design at the same time as procurement and installation. This had not been the original intent of the project and the Project Team had not been resourced accordingly</td></tr>
<tr><td colspan="2">Contract and supplier management</td></tr>
<tr><td colspan="2">What external suppliers are being used?
➤ Main engineering contractor — completed design and supported overall schedule management
➤ Civil contractors
➤ Plant and equipment vendors
➤ Instrument vendors
➤ Electrical and Instrumentation contractors
What is the external supplier status and performance?
Not good — lack of communication, due to lack of project management and control resource, due to parallel completion of design and installation activity</td></tr>
<tr><td colspan="2">Project controls strategy</td></tr>
<tr><td colspan="2">Are project costs under control? (For example review cost plan — actual versus budget)
Project budget was controlled and reported monthly. The possibility of project overspend had not been properly identified by the site Project Team
What is the likelihood that project budget will be maintained? (Forecast to completion?)
Had the full project been completed, the project would have been 25% over budget
Is the project schedule under control? (For example review schedule and milestone progress)
Due to the overstretching of the site Project Team they had little or no input into the project plan or its review. This activity was delegated to the engineering contractor who worked on it in isolation with little or no feedback regarding what took place on site. The plan was not particularly relevant to what took place on site and had little or no impact on the outcome of the project, as it wasn't used</td></tr>
</table>

(continued)

Table 5-8 (Continued)

<table>
<tr><th colspan="2">Project Management Toolkit — 'In Control?' Checklist</th></tr>
<tr><td>Project: Production plant upgrade</td><td>Project Manager: Confidential</td></tr>
<tr><td>Date: Confidential</td><td>Page: 3 of 3</td></tr>
<tr><th colspan="2">Project controls strategy</th></tr>
<tr><td colspan="2">What is the likelihood that project schedule will be achieved? (Forecast to completion?)
This was reviewed infrequently and data was not accurate, due to lack of time spent on this type of activity. Schedule overran
Are there any changes to scope? (Quantity, quality and functionality) Are cost and schedule under control?
Change control was not instigated appropriately and many design changes took place without the prior knowledge of the site Project Team — thus impacting cost and schedule
There was no effective control strategy in place including change control</td></tr>
<tr><th colspan="2">Project review strategy</th></tr>
<tr><td colspan="2">Are regular Stage Three reviews being conducted? (Is performance being managed and monitored?)
This type of review was not completed for the project at any stage in its life
Is project performance adequate for project success?
No — project performance was not adequate and the project was not successful
Is there regular reporting? (Are the Project Team adequately managing communication of progress and performance to all stakeholders?)
Reporting to stakeholders was not completed</td></tr>
<tr><th colspan="2">Stage Three discussion</th></tr>
<tr><td colspan="2">Is the project under control? (Is the project control strategy robust enough for project delivery to continue?)
As soon as the project started it was out of control. Had a formal review taken place at the time of the change to the project, when all of the phasing requirements were known about, mitigation plans could have been put into place and control regained
What is the certainty that the project will be successful?
This project cannot be regarded as successful. Only 50% of the actual project was delivered, although 100% of design was completed. Budgets were exceeded and project completed late. The Project Team were completely de-motivated for much of the project. It is recommended that site project change procedures are re-evaluated in order to identify what a change is and that a more robust process for the development of project delivery plans is instigated for all future projects</td></tr>
</table>

Case study H — using the in control tools to stop a project

This case study demonstrates what can happen when you use the in control tools on a project that is out of control. As a result the project was slowed down until a robust delivery plan was in place.

Product launch

Situation

A new pharmaceutical product has been developed which would be of significant benefit to a large patient group globally. There would be little serious competition due to the efficacy of the product and the way it can be administered compared to conventional treatments. The board were therefore keen to fast track this product launch and had committed to getting it to the market place 18 months faster than usual new product launches.

The project

The project commenced once the product concept had been proved. Since that time the broad team of scientists, engineers, commercial and supply chain personnel had been working in various functional sub-teams to fast track project progress.

The project leader had a background in regulatory affairs as this was seen to be a particularly difficult part of the project; achieving full approval from the appropriate authorities to manufacture and sell the product.

With product launch only 6 months away, the board wanted additional assurance that the launch would be successfully achieved. They were about to confirm the launch date externally — from that point there would be no going back.

The project leader had links into the engineering project management organization and asked if a detailed review could take place.

The outcome

A project review was held and the initial recommendation was that the team be brought together for a formal risk review. Without this the Project Manager conducting the review had refused to comment on the likelihood that the project would be a success. There wasn't enough 'hard' data for the 'hard' message that the Project Manager was starting to put together. The review checklist (Table 5-9) had already highlighted concerns about the level of project control.

The risk review was held (Table 5-10) and a summary critical path of success collated for senior management (Figure 5-13). The key message was that it was highly unlikely that the launch could be successful in 6 months, however with suitable mitigation plans a plan for a launch in 10 months looks highly probably.

The board were taken through the risk data and bought into the recommendations. The launch date was announced to the marketplace and when the launch was achieved on schedule the industry was impressed at the reduced timescale to get the product to the marketplace — over a year earlier than normal!

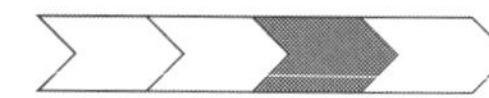

Table 5-9 A completed 'In Control?' Checklist — case study H

Project Management Toolkit — 'In Control?' Checklist

Project: Product launch	**Project Manager:** Confidential
Date: Confidential	**Page:** 1 of 3

Stage Two check

Have there been any changes since Stage Two completion?
There have been significant changes in the supply chain for this new product and the Project Team have not taken these into account. The functional nature of the teams with integration completely in the Project Leader's hands appears to be one problem

Business change management

What is the current status of stakeholder management? (Review the original stakeholder map and discuss)
The plan was to maintain regular contact with senior stakeholders and to get team members to buddy up with key stakeholders in all parts of the supply chain. There seems to be a 'disconnect' as the communications have gone to senior executives assuring them of the product launch to plan, but there has been little discussion with others in the supply chain
A new plan has now been developed by the project leader

How will the business be expected to operate as a result of the completion of the project?
The whole supply chain is being developed for this new product — new raw materials and a new manufacturing process. Both should link in to current distribution processes

Is the business ready for the project?
Current progress-against-change plans are unknown

What is the strategy for handover of the project to the business?
At product launch the project is determined to be complete and ongoing management of all parts of the supply chain will be completely handed over to the business

Scope definition

Has the scope changed since Stage Two completion?
No — the scope remains the launch of a new product which needs to be robustly produced and approved for manufacture and sale by the appropriate regulatory authority

What is the project progress against the Stage Two defined and prioritized project objectives?
Unknown — progress is only currently being monitored against the regulatory strategy as it is so critical to the product launch

Project Roadmap

What project stages/stage gates have been completed? (For example project approved, design complete)
This is a product development project, which also contains a facility revamp project and a supply chain project. The majority of the focus is on the former

(continued)

Table 5-9 (Continued)

Project Management Toolkit — 'In Control?' Checklist	
Project: Product launch	**Project Manager:** Confidential
Date: Confidential	**Page:** 2 of 3
Risk and issue management	
Have all project risks been reviewed regularly during project delivery? No — some of the functional sub-teams have developed informal risk logs but there has been no team review of what will prevent success. Strong recommendation that this occurs as soon as possible **What is the status of mitigation plans?** None **What is the status of contingency plans?** None **What is the overall likelihood of achieving the project CSFs?** It is impossible to comment at this stage other than to raise serious concerns	
Project organization	
Are project activities being completed by the appropriate members of the organization? Yes — however the Project Leader has already raised the issue of the functional silos and a team reorganization is to take place. This will integrate a member of each functional team onto a core team. This should mitigate one of the obvious (if not stated) risks	
Contract and supplier management	
What external suppliers are being used? The only external suppliers are involved with the supply chain and the facility revamp project. The remainder of the capability to complete this project is internal **What is the external supplier status and performance?** The supply chain Functional Leader has mentioned that he is having some issues with the raw materials supplier and he is not yet sure if he can get enough for the full launch stock	

(continued)

Table 5-9 (Continued)

Project Management Toolkit — 'In Control?' Checklist	
Project: Product launch	**Project Manager:** Confidential
Date: Confidential	**Page:** 3 of 3
Project controls strategy	
Are the project costs under control? (For example review cost plan — actual versus budget) Costs do not appear to be an issue here. Because the personnel are all internal and are all already assigned to this critical project it is generally agreed (by the team and by senior management) that cost reporting would not be value-added. The only robust cost control is on the facility revamp project and the supply chain material procurement **What is the likelihood that project budget will be maintained? (Forecast to completion?)** Unknown and not a critical issue for this project **Is the project schedule under control? (For example review schedule and milestone progress)** No — difficult to get a clear picture of the overall schedule. A stage gate process is being used but the problem is that the final two stage gates involve the majority of the work and there is little monitoring in between **What is the likelihood that project schedule will be achieved? (Forecast to completion?)** Unknown — from anecdotal data the suggestion is that there are key items on the critical path which are unlikely to be delivered to plan for example regulatory approval, facility revamp completion and raw materials deliveries **Are there any changes to the scope? (Quantity, quality and functionality) Are cost and schedule under control?** The quality of the 'scientific scope' is robust, as is its link through technical transfer to the manufacturing process and then to the manufacturing facility. However there is no robust schedule control and the impact of changes is not assessed. The schedule is taken as fixed but the targets are not being achieved. The schedule is out of control	
Project review strategy	
Are regular Stage Three reviews being conducted? (Is performance being managed and monitored?) Technical reviews occur on a monthly basis. Regulatory reviews occur each week. These two reviews, plus reporting to senior management and liaison with regulatory authorities, take up the majority of the Project Leader's time. This is a team of senior professionals and their performance is taken for granted **Is there regular reporting? (Is the Project Team adequately managing communication of progress and performance to all stakeholders?)** The Project Leader reports in person to the board at board meetings. Project Team members email the Project Leader when asked	
Stage Three decision	
Is the project under control? (Is the project control strategy robust enough for project delivery to continue?) No — however a risk assessment conducted as a part of this review should demonstrate where action is needed for control to be regained **What is the certainty that the project will be successful?** Success against the current launch date does not appear possible and it is suggested that discussions with senior executives begin soon to move the launch date out by 3 to 6 months. If the date is changed now, well in advance, the impact on the company share price is thought to be low. Data from a risk assessment should support this view and it is suggested that discussions with the board occur after the risk assessment	

Table 5-10 A completed Risk Table — case study H. Note that this is an extract from the Risk Assessment which is several pages in length — approximately 30 risks across the six categories — an example risk in each category is shown

***Project Management Toolkit* — Risk Table**						
Project: Product launch			**Date:** Confidential			
Risk description			**Risk assessment**		**Action planning**	
Risk number	Risk description	Risk consequence	Occur?	Impact?	Mitigation plan	Contingency plan
CSF Level 1 — R&D successfully transfer the technology to full-scale						
1.1	Scale-up problems concerning yield	Exact plant capacity is decreased	Medium	Medium	Amber — work with manufacturing to understand the capacity calculations based on yield and cycle time	Work to decrease cycle time
CSF Level 2 — the clinical trial data proves the required stability						
2.1	Product is not stable under proposed storage conditions	Shelf life has to be decreased	Low	High	Green	Not applicable
CSF Level 3 — the manufacturing facility has been upgraded on schedule						
3.1	New equipment commissioning issues	Impact ability for facility to re-start as scheduled	Medium	High	Amber — engineer to visit the equipment vendor to review operating modes and likely commissioning issues	24-hour working
CSF Level 4 — the approval to make and sell is achieved within the current launch time frame						
4.1	Regulatory package development if delayed	Approval is delayed	High	High	Red — maintain effective dialogue with regulators, action plan for data collection and analysis	Delay launch
CSF Level 5 — launch stock can be manufactured and distributed in the required volumes						
5.1	Raw materials supply issues	Insufficient raw material to make the launch stock	High	High	Red — contact supplier to discuss problem and action plan	Phased launch stock

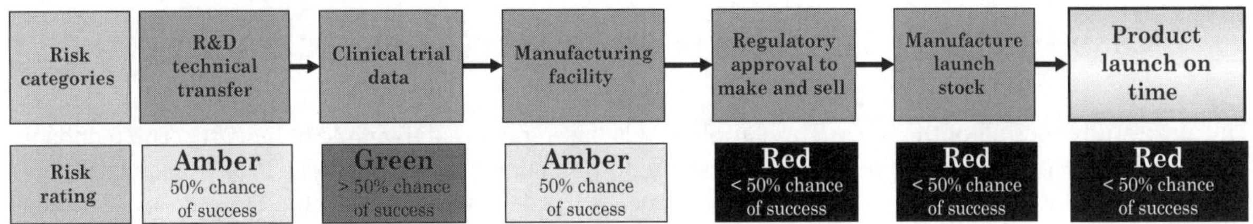

Figure 5-13 An analysed critical path of success — case study H

The critical path of success in Figure 5-13 shows the situation at the end of the first risk assessment which recommended that the launch be put on hold for 3 to 6 months (based on the assessment of mitigation plans which would need to be put in place).

Case study I — using the in control tools to review a project

This case study uses all of the Stage Three tools and is based on a real example in industry. The feedback at the end of the project was 'The project was so smooth I'm not sure we needed a Project Manager, where were all the fires to be put out? Where was all the hassle? I've even got what I asked for!'

New bulk pharmaceutical production facility

Situation

A site was going through a phased expansion requiring a new bulk manufacturing facility and associated infrastructure and support systems.

The first phase of the project had gone badly with cost and schedule spiralling out of control and start-up issues still on-going. The corporate engineering group reviewed the project and concluded that there was no robust control strategy in place:

- Design changes were excessive and design was still in progress when procurement started.
- Equipment vendors and site-based contractors were allowed to 'manage themselves'.
- In order to try to maintain schedule the site was flooded with many different contractors in the final months leading to quality, safety and coordination issues — as well as further project delays.

The second phase was due to commence, and due to the above, a project director from corporate engineering was put in charge. All work was completed by a local site-based Project Manager.

The project

The Project Manager kicked off the second phase of the project and provided a detailed plan of how the project would proceed before he commenced any significant work. The project director approved the plan noting that it outlined a detailed and structured control strategy using appropriate tools.

The project was a traditional engineering one progressing successively through design, procurement, construction and commissioning. Additionally each element needed to take into account the regulatory requirements of the pharmaceutical industry — current good manufacturing practices.

Throughout the project, the project director conducted regular project health-checks and the attached is an example of one completed during final phases of the procurement stage (Table 5-11).

The outcome

The project was completed on time and 10% under-budget. There were no safety or quality issues on site.

Commissioning did need to implement the contingency plan of employing external process engineers but this was easily accommodated within the contingency fund.

The project director was impressed with the overall running of the project and asked for a comprehensive close-out report to be developed for issue to other site engineering departments as a part of sharing learning.

The site director wondered why a project director was needed — everything seemed to go so smoothly (Table 5-12, Figure 5-14 and Table 5-13).

Table 5-11 A completed 'In Control?' Checklist — case study I

<table>
<tr><th colspan="2">Project Management Toolkit — 'In Control?' Checklist</th></tr>
<tr><td>Project: New facility</td><td>Project Manager: Confidential</td></tr>
<tr><td>Date: Week 33</td><td>Page: 1 of 2</td></tr>
<tr><th colspan="2">Stage Two check</th></tr>
<tr><td colspan="2">Have there been any changes since Stage Two completion?
The design change register has been reviewed and confirmed — all now closed-out and design is 100% complete. The system worked well</td></tr>
<tr><th colspan="2">Business change management</th></tr>
<tr><td colspan="2">What is the current status of stakeholder management? (Review the original stakeholder map and discuss)
The Site Director is very happy that the facility will be handed over on time and within budget. This is good public relations for him and his team. The Manufacturing Manager is equally happy with the design of the facility. He didn't get all the automation he requested but upon reflection he can understand the reasons why the dissolution room is to remain essentially manual
How will the business be expected to operate as a result of the completion of the project?
The operators are all used to working on a pharmaceutically regulated manufacturing plant although the new changing regime will be different to current ways of working
Is the business ready for the project?
Operators have been involved in the layout and process flow sheet reviews and understand the basics of the new facility. They will receive training during commissioning. There will be a specific session on the new change regime
What is the strategy for handover of the project to the business?
Process Engineers who will eventually work in the completed facility will be a part of the commissioning team. The handover will take place after the performance of the facility has been proved at the end of commissioning</td></tr>
<tr><th colspan="2">Scope definition</th></tr>
<tr><td colspan="2">Has the scope changed since Stage Two completion?
The basic scope, in terms of plant capacity, quality and operating requirements, has not changed although some issues were caused by the integration of the new change regime which needed to be incorporated into the layout design
What is the project progress against the Stage Two defined and prioritized project objectives?
See attached progress report (Table 5-13)</td></tr>
<tr><th colspan="2">Project Roadmap</th></tr>
<tr><td colspan="2">What project stages/stage gates have been completed? (For example project approved, design complete)
Design — 100%, procurement — 95%, construction — 10%, commissioning — 0%</td></tr>
<tr><th colspan="2">Risk and issue management</th></tr>
<tr><td colspan="2">Have all project risks been reviewed regularly during project delivery?
Yes — a risk review is conducted each month and also at the start or end of a project phase
What is the status of mitigation plans?
Mitigation plans appear to be well managed and have ensured that no risks have yet converted to live project issues
What is the status of contingency plans?
No contingency plans have been used although a few were developed in the early stages of the project when design progress appeared to be an issue
What is the overall likelihood of achieving the project CSFs?
There is a high probability that the project will meet its CSFs</td></tr>
</table>

(continued)

Table 5-11 (Continued)

<table>
<tr><th colspan="2">Project Management Toolkit — 'In Control?' Checklist</th></tr>
<tr><td>Project: New facility</td><td>Project Manager: Confidential</td></tr>
<tr><td>Date: Confidential</td><td>Page: 2 of 2</td></tr>
<tr><th colspan="2">Project organization</th></tr>
<tr><td colspan="2">Are project activities being completed by the appropriate members of the organization?
Yes — following an initial period of team confusion when the engineering contractor had problems maintaining a consistent team. The internal team is small — Project Manager and Process Engineers. The role of the Project Director has been made easy by the effective project management practices already in place</td></tr>
<tr><th colspan="2">Contract and supplier management</th></tr>
<tr><td colspan="2">What external suppliers are being used?
➤ Engineering contractor — design contract and management of procurement, construction and commissioning
➤ Sub contractors for site works
➤ Equipment vendors
What is the external supplier status and performance?
Initially equipment vendors were not managing the contracts robustly and it looked like deliveries would slip but regular visits and an understanding of the issues within overseas vendors in particular has resolved this risk</td></tr>
<tr><th colspan="2">Project controls strategy</th></tr>
<tr><td colspan="2">Are project costs under control? (For example review cost plan — actual versus budget)
Yes — cost plan is detailed and contingency has been calculated based on an assessment of risk. Contingency run-down is being monitored and an earned value overview has ascertained that engineering man-hours are being spent more efficiently than planned (Table 5-12). A robust contract strategy with appropriate contract terms has ensured that capital is also being spent efficiently
What is the likelihood that project budget will be maintained? (Forecast to completion?)
High probability of significant savings —up to 10%
Is the project schedule under control? (For example review schedule and milestone progress)
Yes — confirmed using a combination of logic linked schedules (overall project, and commissioning), milestone (procurement), earned value (design and engineering management and also construction)
What is the likelihood that project schedule will be achieved? (Forecast to completion?)
High probability of completion on time
Are there any changes to scope? (Quantity, quality and functionality) Are cost and schedule under control?
Yes — a contract variation control system is in place as well as the design control system
Contract variations constitute 25% of contingency fund to date (better than forecast at 35%)</td></tr>
<tr><th colspan="2">Project review strategy</th></tr>
<tr><td colspan="2">Are regular Stage Three reviews being conducted? (Is performance being managed and monitored?)
Yes — monthly review with the project director, monthly reviews with the engineering contractor, weekly design team meetings (now stopped), procurement updates every 2 weeks, weekly site construction review
Is there regular reporting? (Are the Project Team adequately managing communication of progress and performance to all stakeholders?)
Yes — detailed report for internal review of progress, one-page summary for team notice board and Project Director (Table 5-13) and overviews based on this one-page report for senior management</td></tr>
<tr><th colspan="2">Stage Three decision</th></tr>
<tr><td colspan="2">Is the project under control? (Is the project control strategy robust enough for project delivery to continue?)
Yes — a robust and appropriate strategy is in use for a project of this size
What is the certainty that the project will be successful?
High probability of success</td></tr>
</table>

Table 5-12 A completed Earned Value Tool — case study I

<table>
<tr><th colspan="4">Project Management Toolkit — Earned Value Tool</th></tr>
<tr><td colspan="2">Project: New facility</td><td colspan="2">Date: Week 33</td></tr>
<tr><th colspan="4">Earned value basis</th></tr>
<tr><td colspan="4">The engineering management contractor total fee is to be monitored by earned value. The contractor provides engineering and management resource for the whole project and represents approximately 22% of the total project costs</td></tr>
<tr><td>Cost management basis</td><td colspan="3">A target-reimbursable basis. Complex series of deliverables over many different engineering functions covering design, procurement, construction and commissioning</td></tr>
<tr><td>Progress management basis</td><td colspan="3">Each function has its own earned value system so that the progress measurement here is a collation of other progress measurements based on deliverable achievement only</td></tr>
<tr><td>Earned value calculation</td><td colspan="3">The progress measurements are all converted to a percentage. This could be converted to a cost via comparison with the target fee agreed. However it was thought to be easier to convert the plan to a percentage completion. Actual costs are based on invoices, which are based on timesheets. This has also been converted to a percentage through comparison with the total cost</td></tr>
<tr><th colspan="4">Earned value data collection (all figures are cumulative)</th></tr>
<tr><th>Timeline</th><th>Plan</th><th>Actual</th><th>Earned value</th></tr>
<tr><td colspan="4">Note that the data table is lengthy for this example and contained on a spreadsheet — the 'S' curve is automatically produced by the spreadsheet (Figure 5-14)</td></tr>
<tr><th colspan="4">Earned value interpretation</th></tr>
<tr><td colspan="4">See Figure 5-14. Initially the contractor worked inefficiently as he suffered from numerous team changes. This also impacted the schedule and an early delay of 1 week was seen. This delay remained with the project until week 27 although the project is presently a week ahead of schedule. Currently the contractor is showing the efficiencies of maintaining the same team and is forecasting that they will complete on time and approximately 4% under budget (less total man-hours needed to complete the work)</td></tr>
</table>

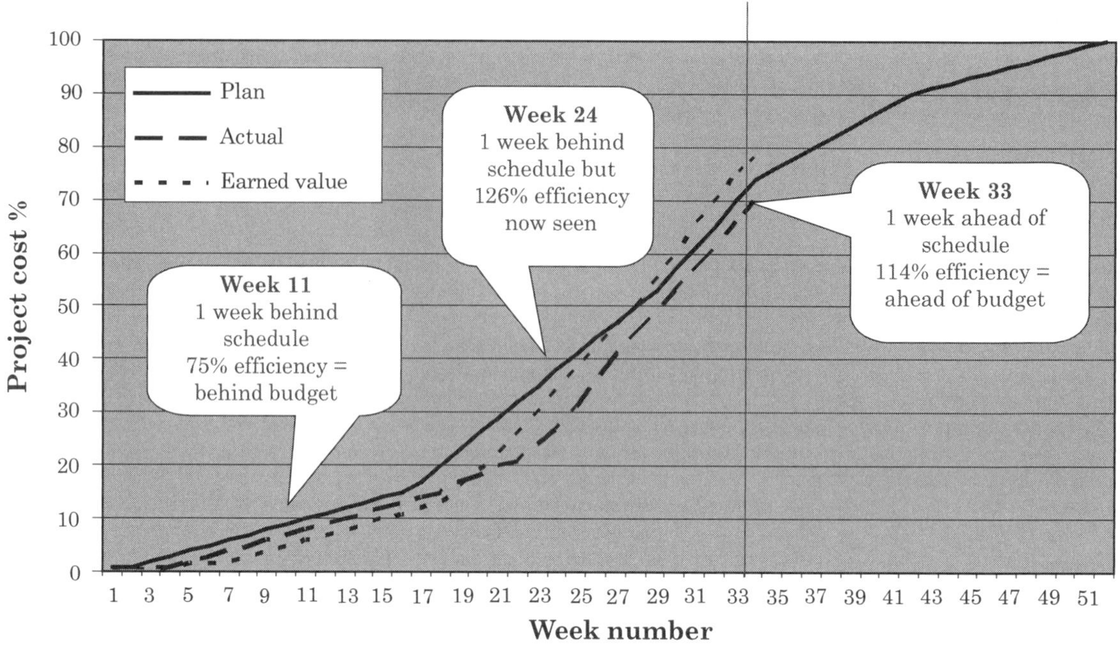

Figure 5-14 'S' curve for case study I

Table 5-13 A completed Project Scorecard — case study I

Project Management Toolkit — Project Scorecard

Project: New facility		**Date:** Week 33	
Project status summary			
Project status	Design is complete, procurement is 95% complete and construction is now well established		
Project control status	Overall cost is forecast to be under budget — 4% saving in engineering contractor fees and approximately 5% saving in capital. However, the capital savings anticipated will remain in the contingency budget until the sub-contracts are nearing completion Overall the schedule is forecast to be achieved. The late start to the civil contract can be mitigated by an early start by the commissioning team The risk profile is amber, which is to be expected at this stage (entering construction). No red risks have been identified at this stage		

Key activities in this period	**Achievements**
➤ The procurement of the final major sub-contracts has been placed ahead of schedule (mechanical and instrument and electrical) ➤ Start of the civil sub-contract site work ➤ Confirmation of all equipment deliveries with vendors ➤ Site safety inspection passed	➤ Significant cost savings negotiated with the mechanical sub-contractor through a measured term style contract with bonuses for completion on schedule ➤ Design 100% complete. The engineering contractor has also confirmed that the team will be available for commissioning if we cannot find process engineers
Key activities in the next period	**Critical issues**
➤ Long-lead equipment items are to be tested prior to delivery to site (to be witnessed) ➤ Final sub-contracts to be placed (insulation, painting and signage)	➤ None at this time

Critical success factor tracking

CSF Level 1	**Progress**		**Risk Rating**	
Technically acceptable design in line with pharmaceutical needs	Green	Design is now 100% complete Scope in terms of cost, quality and functionality has been achieved	Green	No risks — design is complete
Procurement of equipment and subcontracts to meet quality, cost and schedule needs	Green	100% of major equipment purchased within 95% of budget 95% of sub-contracts placed Final minor items need to be purchased within the month	Amber	The potential for cost variations has been minimized through the development of bonus based contracts but this needs to be tracked Overseas equipment issues have been eliminated

(continued)

Table 5-13 (Continued)

***Project Management Toolkit* — Project Scorecard**				
Project: New facility			**Date:** Week 33	
Critical success factor tracking				
CSF Level 1	**Progress**		**Risk rating**	
Construction technically acceptable, safe and meets quality, cost and schedule needs	Amber	The civil site work commenced a week behind schedule due to late placement of sub-contract due to extended negotiation and issues with the design	Amber	There are many potential construction risks and the risk rating remains amber until all construction activities on the critical path have commenced Earned value tracking of pipe-work meterage and electrical and instrumentation cabling is to be used
Commissioning	Green	Commissioning included in equipment and sub-contract scope but yet to commence	Amber	Still a concern that there may not be enough internal process engineering resource to support the commissioning — see above
Overall project	Amber	Slight construction delay seen but expected to be resolved and overall forecast to achieve project schedule	Amber	Overall risk rating remains amber — all mitigation plans are working but tracking should continue

Handy hints

Never be afraid to control!

You should never be afraid to check how the project is doing versus your plan. If you don't check then the chances are you are already deviating from your plan. The longer you leave it the more you will deviate and the more out of control you will be.

Monitoring is not control

Project Managers are focused on delivery however it is easy for the inexperienced Project Manager to cycle out of control by collecting the mass of project data that is thrust at him on a daily basis. Alternatively even the experienced Project Manager can get bogged down by the generation of the dreaded monthly report. If you don't analyse the data you're not in control — you're just monitoring history!

Keep your sponsor involved

Many projects have failed to deliver what was expected of them because of a lack of organizational support. If your sponsor understands your progress he is confirming, on an on-going basis, that it aligns with the business benefits, which he is accountable for delivering.

Don't forget to 'see the wood for the trees'

It is very easy to get sucked into the detail when you are managing a project or to fall back and manage at a micro-level the technical areas you know best — you have to pull yourself back and check that the overall project is in control and will deliver success.

Always look to expand your Stage Three toolkit

The Stage Three toolkit contains four tools out of many that are actually used in the delivery stages of a project. There are other tools that will help you as you delivery your project — search them out and use them appropriately and pragmatically.

Further reading

Project delivery is a huge subject and this chapter has only introduced a few of the core topics. These aspects are covered in more detail by *Managing Project Delivery: Managing Control and Achieving Success* a further book in this Project Management Series (ISBN: 978-0-75068515-3).

Additionally the websites referenced on page xi can be used to source further 'best practice' material.

And finally . . .

- Ask 'are we in control?' — if you don't know then the chances are you're not!
- Understand the value of appropriate value-added control tools and techniques. Don't let the control tools become the project — be pragmatic!
- Be flexible — use and adapt the tools to achieve your goals.

6 Stage Four: benefits realized?

The final value-added stage in a project involves asking 'have we delivered the benefits?' However, as this stage is more often then not outside the scope of a Project Manager's role — it doesn't get asked as often as it should.

In Chapter 3 the benefits management life-cycle was introduced (Figure 3-1 on page 13) — a continuous linkage between the project and the business throughout the life of the project. Stage Four of a project is confirming that this linkage has been successful and the reasons why the project was needed have been sustainably delivered.

Benefits realization (Figure 6-1) is concerned with the:

- Tracking of benefits delivery after the project scope has been delivered.

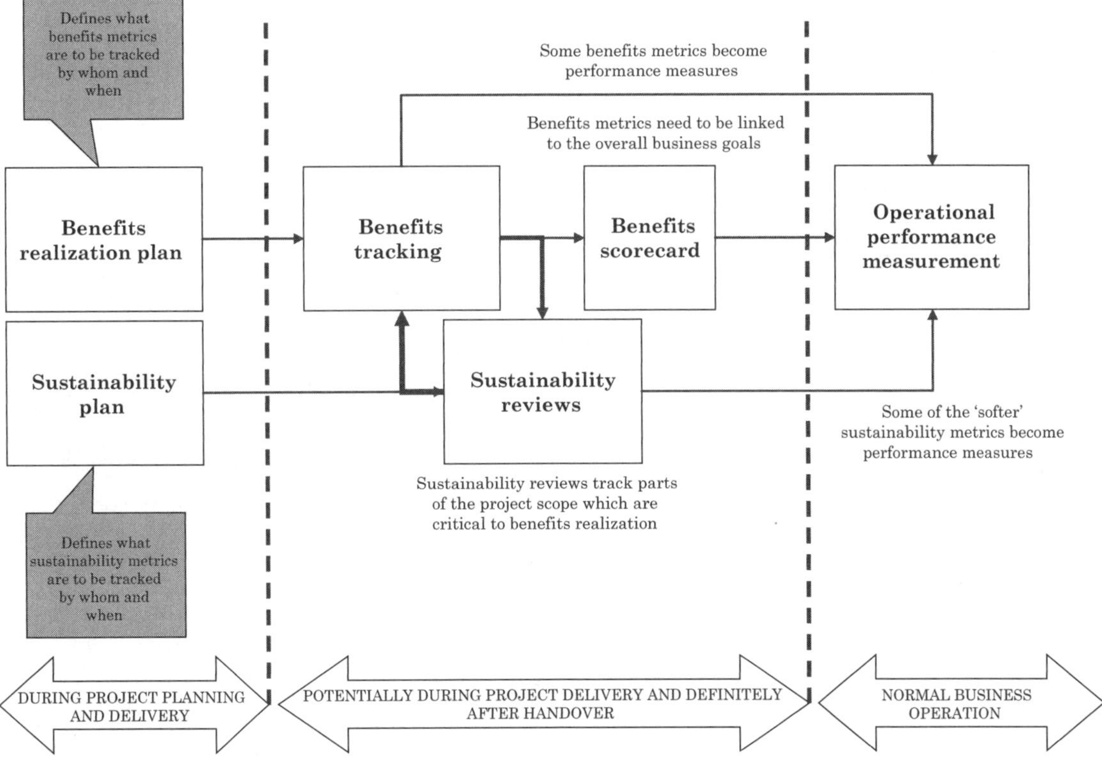

Figure 6-1 The benefits realization process

- Assessing if the achievement of the project objectives will allow the realization of the improvement in specific benefit metrics.

The key to benefits realization is twofold:

- Ownership — who is accountable for the delivery of the benefits?
- Benefits tracking — how, who and when the benefit metrics are tracked.

The key to both of the above is understanding sustainability.

For some types of project the benefits are delivered as the project is being delivered, for example business improvement projects; for others the benefits can only be realized when the full project scope has been delivered, for example launching a new product or building a new production facility. However, what is common in all projects post-delivery is the need to check the sustainability of the solution to the business need.

Sustainability

Sustainability is the act of integrating any business change within the organization so that it becomes the normal way of working — 'the way we do things around here' — examples are shown in Table 6-1. Note that some elements of sustainability are within the control of the organization whereas others are not.

Sustainability can be checked after the project has been completed through a series of reviews, which are specific to the project. Sustainability, like benefits realization, starts during the early phases of a project — it must be planned for and it must be built into the project scope that is delivered. Any areas outside of the scope of the project need to be identified at an early stage in the project's life.

The following tools are just a few of those which can be used during the final stages of a project to provide answers to 'have the benefits been sustainably realized?'

- Benefits Realized Checklist.
- Benefits Tracking Tool.
- Project Assessment Tool.
- Sustainability Checklist.

Projects, which do not receive this level of challenge at this late stage in their life, cannot really demonstrate that they have delivered a benefit to the business and therefore were successful.

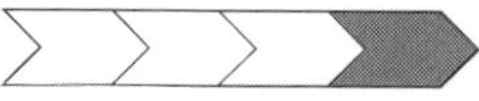

Table 6-1 Examples of sustainable benefits

Project type	Project scope	Benefits realized during project delivery	Benefits realized after project delivery	Requirement for sustainability
Engineering improvement	Identify equipment and procedures to be changed Implement changes and train personnel	Procedures may change as the project advances and performance improvements are seen	Performance improvements within the specific areas of the plant	Procedures need to be followed and equipment maintained as designed
New manufacturing facility	Design and build a manufacturing facility	None	Manufacturing facility operates in line with business needs (producing product at the required cost) and that product is sold at the forecast price	Margin maintained (selling price minus operating cost) Facility is operated and maintained in line with the design
Product launch	Develop the product and process, the manufacturing, regulatory and supply chain strategies and the marketing plan	None	Supply chain delivers the new product to the market place The company achieves the expected market share and expected market price	Supply chain operates as designed Process and product is technically robust Market share and margin maintained
Organizational development	Design a new, more appropriate way of working	Costs, quality and process lead-time can all change during the project — sometimes dipping before improving	The finalized organization design should demonstrate the cost, quality or lead-time improvements	The organization continues to operate as designed People do not revert back to 'old ways'

Tool: 'Benefits Realized?' Checklist

If any project (during the latter stages) cannot be challenged by the questions contained in this checklist *and* have robust answers, suitably backed up, then the project is unlikely to be able to support the sustainable delivery of the business benefits.

This tool can be used at the end of Stage Three (at project handover to the business) and then at any point in Stage Four as an ongoing 'health check' or as part of a sustainability review.

The 'Benefits Realized?' Checklist explained

The checklist contains five major checks:

- Stage Three check.
- Business benefits.
- Business change.
- Scope definition.
- Stage Four decision.

Table 6-2 shows the 'Benefits Realized?' Checklist with high-level guidance on how to complete each check. The following are additional, more detailed, notes to support checklist completion.

Stage Three check

The intent of this check is to confirm that this is an appropriate time to start to check the realization of the benefits. This review can be linked to the final health check conducted during the project delivery; using the 'In Control?' Checklist (see page 109):

- Have there been any changes since Stage Three completion?

Typically this checklist is performed on a 'live' project at a stage when the sponsor believes that the benefits are being realized. In this way it can act as a formal close-out between the sponsor and the customer. It is strongly linked to sustainability reviews.

Business benefits

These four checks close the loop on the benefits management life-cycle originally introduced in Chapter 3:

- Has the business case changed since Stage One?
- Have all benefits been defined in terms of trackable metrics?
- What is the stakeholder feedback?
- Are the benefits being tracked?

The business case represents the agreement on what the project needed to enable the business to deliver and any changes should have been formally noted and agreed by all appropriate stakeholders.

The benefit metrics are the specific metrics, which have been tracked to demonstrate that the business case has been delivered. These will have been progressively defined from Stage One through Stages Two and Three and then tracked in Stages Three and Four as appropriate to the specific project.

Table 6-2 The 'Benefits Realized?' Checklist explained

***Project Management Toolkit* — 'Benefits Realized?' Checklist**	
Project: *<insert project title>*	**Project Manager:** *<insert name>*
Date: *<insert date>*	**Page:** 1 of 1
Stage Three check	
Have there been any changes since Stage Three completion? (Note only the changes since the final Stage Three 'health check') *<insert comments regarding any changes to the project since the previous health check>*	
Business benefits	
Has the business case changed since Stage One? (For example during planning and delivery, pre- or post-project approval) *<review the formal business case and confirm its validity — insert any additional comments if there have been changes since approval — note if formal change agreements were made with the sponsor>* **Have all benefits been defined in terms of trackable metrics? (Why is the project being done?)** *<insert comments with regard to the status of the Benefits Specification Table and whether this has been converted into a Benefits Tracking Table>* **What is the customer feedback? (Feedback from all stakeholders in the business including the customer)** *<insert comments regarding any feedback from the customer particularly relating to the benefits delivery>* **Are the benefits being tracked?** *<attach a copy of the Benefits Tracking Table — comment on how long the benefits have been tracked>*	
Business change	
Is the business ready for the project? (If the project can only enable benefits delivery by changing the way people work — has this been delivered, for example training?) *<consider how the business has reacted to the changes delivered by the project — attach any completed sustainability reviews>*	
Scope definition	
Has the scope been delivered? *<insert comments on the delivery of the project as agreed — refer to the CSFs>* **Have the benefit enablers been delivered? (Are you sure that the project will enable the benefits to be delivered now the project is complete?)** *<consider if the delivered project enables the benefits to be delivered as agreed by the business case>*	
Stage Four decision	
Has the project been delivered? (Delivery of project critical success criteria) *<insert the decision — yes or no — with back-up material such as a completed after action review>* **Have the business benefits been delivered? (Why was the project done in the first place?)** *<insert the decision — yes or no — with back-up material such as a completed benefits scorecard>*	

The final review in terms of benefits relates to the views of the various stakeholder groups, for example the customer, senior business stakeholders, and user groups who are impacted by the project. Their view on the delivery of the benefits is likely to be based on 'soft' data — this is entirely valid and will have an impact on the sustainability of the changes that the project has delivered.

Business change

As stated in Chapter 3 — every project has the potential to change the nature of the business or organization into which it is being delivered. This check therefore challenges whether these changes have been appropriately managed and also whether additional business changes, necessary for the benefits realization, have been delivered by the business:

- Is the business ready for the project?

A typical business change required by a project could be organizational changes — for example the project can only support the delivery of the benefits if the people within the business are re-organized or re-trained.

This check therefore ensures that the sponsor has delivered one end of the bargain — supporting the business in identifying the changes which need to be made; and the customer has delivered the other — making the changes so that the business is ready for the completed project.

Scope definition

These checks are challenging whether the scope that the project has delivered will enable the benefits to be realized:

- Has the scope been delivered?
- Have the benefit enablers been delivered?

During the delivery of a project, changes occur for a variety of reasons and although a project change control process should address the issue of benefits delivery (see page 8) not all change control procedures are this robust.

Stage Four decision

Once all the available project information has been gathered it is critical that the Project Manager, in partnership with the project sponsor and the customer, makes a clear decision on the success of the project by asking:

- Has the project been delivered?
- Have the business benefits been delivered?

The end point for Stage Four is an agreement that the project was a success in terms of both its internal ways of working and also its organizational impact:

- The effective management of the project enabling the delivery of the Critical Success Factors (CSFs).
- Effective benefits management — delivery of 'why we did the project in the first place'.

Using the 'Benefits Realized?' Checklist

This tool can be used in different ways:

- To formally check the delivery of benefits once a project has been completed and the change sustained.
- To check the progress of benefits realization prior to confirming complete sustainability of the changes introduced by the project.
- To trouble-shoot when a project under-delivers on anticipated benefits.

In each mode the checklist is always best completed in partnership with the customer and the sponsor — the customer needs the business benefits and the sponsor is accountable for their delivery. Often the Project Manager is best placed to facilitate this session and to use the checklist as prompts in the discussion. The ultimate aim of the session is to answer one question:

- Has the project been successful? (Stage Four decision) Success being the delivery of the business benefits as well as the project objectives.

To complete the checklist both the customer and the sponsor will need access to the data that supports the demonstration of project success:

- 'Hard' data such as the trackable benefit metrics, for example cost savings and performance improvements.
- 'Soft' data such as stakeholder feedback and other measures used to check the sustainability potential of the project outcome.

Each of the case studies in this chapter uses the 'Benefits Realized?' Checklist to demonstrate the above.

In completing the checklist it can be seen how the use of other tools within this chapter can support the demonstration of the delivery of the anticipated business benefits and therefore the completion of the Stage Four decision.

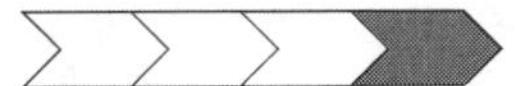

Tool: Benefits Tracking Tool

The aim of this tool is to confirm the realization of the benefits from a project. The completion of this tool supports positive responses to the checks in the 'Benefits Realized?' Checklist (see page 152).

The Benefits Tracking Tool is based on standard tracking techniques — comparison of an actual metric with a planned metric at a specific milestone date or after completion of a specific milestone activity. It is in effect the project benefits realization plan:

- A time-line showing how the benefit metrics should actually be realized.

The Benefits Tracking Tool explained

This tool contains four main sections and is an extension of the Benefits Specification Table introduced in Chapter 3 (see page 26):

- Benefit metric.
- Baseline.
- Milestone.
- Target.

Table 6-3 shows the Benefits Tracking Tool with high-level guidance on how to complete each section. The following are additional notes to support tool completion.

Table 6-3 The Benefits Tracking Tool explained

***Project Management Toolkit* — Benefits Tracking Tool**

Project: *<insert project title>*				**Date:** *<insert date>*		
Benefit metric		**Baseline**	**Milestone 1**	**Milestone 2**	**Milestone x**	**Target**
		<insert baseline date>	*<insert date or activity>*			*<insert target date or activity>*
Metric 1 *<insert metric from Benefits Specification Table>*	Plan	*<insert baseline data>*	*<insert planned metric level>*			*<insert target metric level>*
	Actual	*<insert baseline data>*	*<insert actual metric level>*			*<insert actual metric level>*
Metric 2	Plan					
	Actual					
Metric *x*	Plan					
	Actual					

Benefit metric

The previously identified benefit metrics need to be reviewed. It is not unusual for the benefit metrics to be better defined by this point — as further data is collected during the project delivery stage.

Baseline

This is the starting point in terms of benefits tracking; ideally it is the level of the benefits metric prior to the project start. If a benefit metric does not have a baseline level then sufficient data must be collected as the project progresses.

Milestone

The realization plan will have identified either specific dates or project activity completion milestones which will trigger a change in the benefit metric level. As these milestones are achieved the actual metric measure is taken and the tracking tool is updated.

Target

This is the end point in terms of benefits tracking. The target benefit metric directly links to the original rationale for the development of the project and the approved business case. If it is not achieved at the planned final milestone then the project has not successfully delivered fully the benefits that were intended.

Using the Benefits Tracking Tool

The Benefits Tracking Tool is a management and control technique to focus on the delivery of the benefits realization plan. The Benefits Tracking Tool takes the base information from the Benefits Specification Table and converts it into a benefits realization plan:

- Base data — the baseline data and the required level that each identified benefit metric must achieve.
- Benefits realization plan — a plan of target metric levels against milestone dates or milestone activities.

It is not usual for a Project Manager outside of business change projects to be involved with this. However, this doesn't mean that it won't be of any use in other types of projects for both project and business managers.

The usefulness of the tool is entirely linked to the accuracy and precision of the data collected. Therefore the data must be reviewed against these criteria. Note that some 'soft' measures are unlikely to give more than a general trend due to the intangibility of the metric.

Figure 6-2 shows how the data from the Benefits Tracking Tool can be converted into graphical trending charts which make both interpretation and communication easier. The example is taken from an improvement project originally introduced in Chapter 3 — a cleaning improvement project: to increase the efficiency of cleaning whilst also reducing the annual site cleaning costs. The benefits from this project, as partially demonstrated by the graphs, were realized:

- The site culture was changed and overall site tidiness did improve — things didn't get as dirty thus reducing the amount of cleaning required.
- The new cleaning contractor did achieve the appropriate benchmark standard of 'clean' — the new contractor came on board during project month four and their improvement trend followed the trend in improvement in the site 'clean' culture.

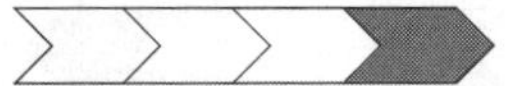

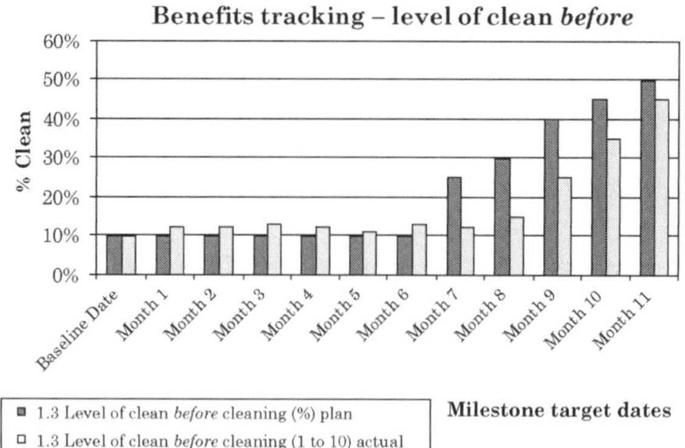

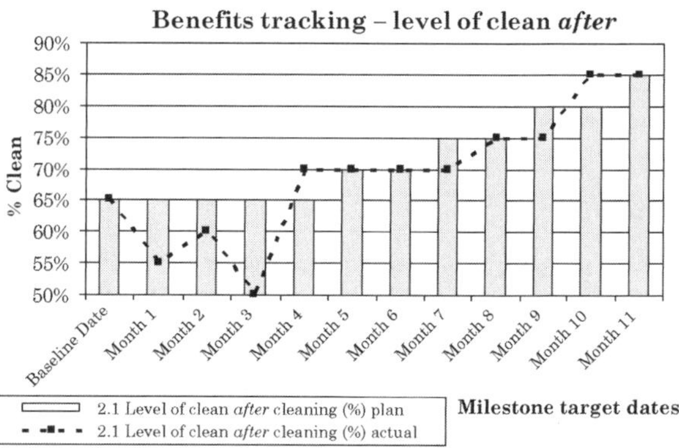

Figure 6-2 Benefits tracking example

The Benefits Tracking Tool is usually owned by the sponsor who is accountable for the realization of the benefits from the project. However, the data would usually be collected by end users to some extent.

Regular reviews between the sponsor and the customer are recommended bearing in mind that this data will support agreement that the changes made by the project have delivered sustainable benefits. Following agreement that the sustainability of the benefits has been achieved, the sponsor hands the Benefits Tracking Tool to the customer for their use.

Often after this point is reached the customer will select some of the benefit metrics as operational measures which will continue to support the effective management of an area within the organization, for example, in the facility cleaning improvement project (Figure 6-2):

- The 'level of clean after' benefit metric is a good indication of the efficiency of the cleaning and is therefore a good measure of an external contractor's ability to meet the appropriate performance level — this was chosen as an operational measure and monitored on a quarterly basis.
- The 'level of clean before' benefit metric was an indication of whether the people working on the site had changed their attitudes to tidiness on the site, and therefore their actions — this was not chosen as an operational measure on the basis that by the time the change had been deemed to be sustainable in month 16 the benefit metric had been stable at the target level for 5 months.

Benefits tracking reporting — the benefits scorecard

The Benefits Tracking Tool summarizes the benefits realized from individual projects, however, within an organization there is usually a need to collate all the benefits data from all projects to understand what overall organizational benefits are being delivered.

A key role for the sponsor is to regularly report benefits from individual projects so that this overall summary can be made. This is typically done via some form of benefits scorecard.

The aim of a scorecard approach is to allow a coordinated approach to the collation of high-level benefit metrics which are then used to make key decisions which impact the future of the business.

By focusing on a few key measures, senior management are able to identify the areas that need to improve if the overall strategic objectives are to be achieved.

The benefits scorecard approach also aims to give strategic 'balance' to the business by looking at benefit metrics in four categories:

- Customer relationship — benefit metrics that demonstrate that an organization's customers are being satisfied.

- Organizational development — benefit metrics that demonstrate organizational growth through knowledge management and individual and team performance.
- Organizational activities — benefit metrics that demonstrate improvement of operational processes.
- Operational effectiveness — benefit metrics that demonstrate an improvement in organizational value.

The benefits scorecard (Figure 6-3) is a reporting format to collate many Benefits Tracking Tools into one succinct summary for an organization or part of an organization.

Similar scorecard formats are used to collate operational measures within the business. However the format shown in Figure 6-3 is focused on what the portfolio of projects within an organization or business area are delivering in terms of sustainable benefits.

Based on the benefits scorecard, senior management, in liaison with sponsors and internal customers, can review if the selected projects are delivering the appropriately balanced benefits as needed by the business.

Usually the scorecard would be backed up by a high-level organizational benefits map. Benefits mapping at a project level is discussed in Chapter 3 and the process outlined there is applicable to the development of an organization benefits map.

The Benefits Tracking Tool is therefore appropriate input data for both the project specific sustainability reviews and the overall project portfolio reviews.

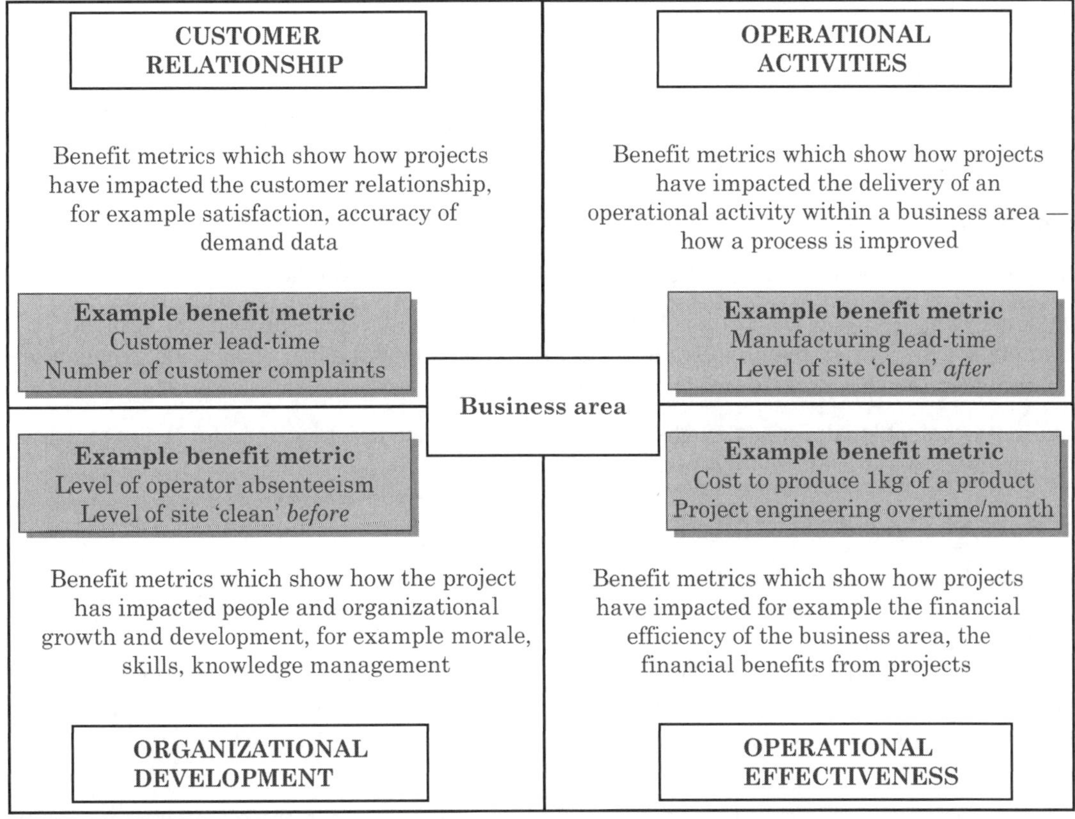

Figure 6-3 The benefits scorecard

Tool: Project Assessment Tool

The aim of this tool is to support the completion of the final project assessment. The completion of this tool supports positive responses to the checks in the 'Benefits Realized?' Checklist (see page 152) however the ultimate aim of this tool is to share knowledge.

The Project Assessment Tool is based on standard techniques of reviewing an activity after its completion to ensure that appropriate actions are taken — an after action review.

The Project Assessment Tool explained

The tool has five separate areas:

- Objectives review.
- Journey review.
- 'What went well' review.
- 'What could have gone better' review.
- Project summary.

Table 6-4 shows the Project Assessment Tool with high-level guidance on how to complete each column. The following are additional, more detailed, notes to support tool completion.

Objectives review

It is always useful to reflect back to the original CSFs. The final Table of Critical Success Factors (see page 70) is useful data for this part of the tool.

Journey review

The key part of this review is the identification of the project 'highs' and 'lows':

- Example 'highs' — when team motivation was good, when a particular problem was solved.
- Example 'lows' — when team motivation was low, when a mistake had been made, when an apparently unsolvable problem was becoming a blocker to the project.

From these project 'peaks' and 'troughs' significant knowledge can be gained for the organization, for the individuals and for future projects.

'What went well' review

Often a Project Team will reflect on the 'good things' without really understanding why that activity or action worked so well. If an organization wants to learn from 'good practice' then it needs to understand the root cause of the 'good practice'.

The five whys analysis can be used here to understand the root cause so that appropriate knowledge is shared (Table 6-5). The team may have taken from the 'good team working' a sense that it was just the team mix that worked well when in fact the real reason the team working was so effective was:

- Robust planning which ensured that the team understood the project, the intended outcome and each other's roles as well as their own.

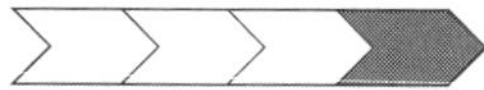

Table 6-4 The Project Assessment Tool explained

Project Management Toolkit — Project Assessment Tool		
Project: *<insert project title>*	**Date:** *<insert date>*	
Objectives review		
Objective/CSF	**Achievement**	**Action (sharing or learning)**
< list the original project objectives/CSFs>	*<note the project achievements against each specific objectives/CSF>*	*<note any specific actions related to the review whether this is to share with a specific group or a learning action>*
Journey review		
<attach a flowchart which describes the journey the project, Project Team and stakeholder have taken>		
Highs	**Sharing**	**Learning**
<note the particular highs>	*<insert sharing actions>*	*<insert learning actions>*
Lows	**Sharing**	**Learning**
<note the particular lows>	*<insert sharing actions>*	*<insert learning actions>*
'What went well' review		
What went well?	**Why?**	**Sharing action**
<insert the categories of areas or activities that went well during the project>	*<conduct a five whys analysis on each category>*	*<decide if there is any knowledge to share>*
'What could have gone better' review		
What could have gone better?	**Why?**	**Learning action**
<insert the categories of areas or activities that could have gone better during the project>	*<conduct a five whys analysis on each category>*	*<decide if there are any learning actions>*
Project summary		
<insert a summary of the project from the Project Team and stakeholder perspective> *<insert a quote reflecting the teams' views>*		

The knowledge shared in this example was the power of the project delivery plan in dealing with the 'softer' side of project management at an early stage in the project. Project Team building is an often neglected or misunderstood activity. It can also be done inappropriately.

'What could have gone better' review

Often a Project Team will reflect on the 'bad things' without really understanding why that activity or action went so badly. If an organization wants to learn from 'poor practice' then it needs to understand the root cause of the 'poor practice'.

The five whys analysis can be used here to understand the root cause so that appropriate lessons can be learnt (Table 6-6). The team may have taken from the 'rushing for deadlines' that they had a personal time management issue when the problem was much more holistic than that:

- At project kick-off the customer had insisted that they be allowed to comment on the majority of deliverables however the Project Manager didn't robustly incorporate this into the plan. As a result the customers received their request for comments late.

Table 6-5 'What went well' review five whys analysis

Question	Answer
Identify 'what went well'	The team worked well together
Why? 1	We all knew what we were doing
Why? 2	We had a detailed RACI Chart
Why? 3	We got together at the start of the project to sort roles and responsibilities out
Why? 4	We were told to (by the Project Manager)
Why? 5 — the root cause	It was in the project delivery plan

Table 6-6 'What could have gone better' five whys analysis

Question	Answer
Identify 'what could have gone better'	We always seemed to have to burn the midnight oil just before a deadline
Why? 1	No matter how ready we thought we were, the final completion actions/time caught us by surprise
Why? 2	There were always last minute changes to key deliverables
Why? 3	The customer comments had to be included but we never get them early enough
Why? 4	We only get them in line with our original requests (a little late) and that is too late for our final work
Why? 5 — the root cause	The project plan didn't incorporate the integration of the customer comments into the workflow and so the customer comments were not linked to the activity they were commenting on

The lesson learnt in this example is that all activities, which can impact the progress of your key deliverables, must be identified — only then can progress be managed and controlled.

Project summary

When projects are delivered and teams disbanded there is always a flavour of 'how the project went' left within the organization. This section aims to capture this as valid knowledge to share and learn from. Example quotes from other projects:

- 'It was like climbing a really big mountain — great once you got to the top!'
- 'I felt like I was on a big dipper in the dark — never knowing when the next dip will happen — bit scary'
- 'It all seemed so effortless — one step followed the next — amazing!'

The above quotes are all metaphors although this isn't a pre-requisite for a valid quote. Sometimes people find it easier to express their feelings (a 'soft' project issue) by referencing other images which made them feel the same. The second example above uses a big dipper fairground ride as an image of something to be scared of filled with uncertainties; another person could have used the same image to describe a great project which exhilarated them because they love big dippers and find the management of uncertainty a thrill.

Using the Project Assessment Tool

The only way to generate the data for the Project Assessment Tool is to conduct an after action review (AAR) and the only way to conduct an AAR is with a team.

Often two different AAR sessions will be conducted after a project is completed:

- Internal — with the Project Team.
- External — with the sponsor and external stakeholders (customers, users).

Both sessions are run in exactly the same way as described in Figure 6-4.

Hints for facilitating such a session are:

- Don't let the Project Manager facilitate — his input is wanted too.
- Do allow enough time for the review — an hour is just not enough.
- At all stages encourage an element of brainstorming (with Post-it® notes, for example) and then a converging session (cluster the Post-it® notes into categories).
- Keep track of all 'sharing actions' and all 'learning actions' and ensure that before the session closes that each have assigned names and dates — assign the tracking of the actions to one person in the session.
- A complete Project Assessment Tool may take any number of pages. Summarizing this into one page is useful. Figure 6-5 is an example summary sheet from a relatively small project.
- An AAR is only complete when all the actions have been completed.

The completed Project Assessment Tool is good back-up data for the Benefits Realized Checklist completion, as it provides both 'hard' and 'soft' data on the success of the project. It is also a good link into the sustainability reviews between the sponsor and the customer.

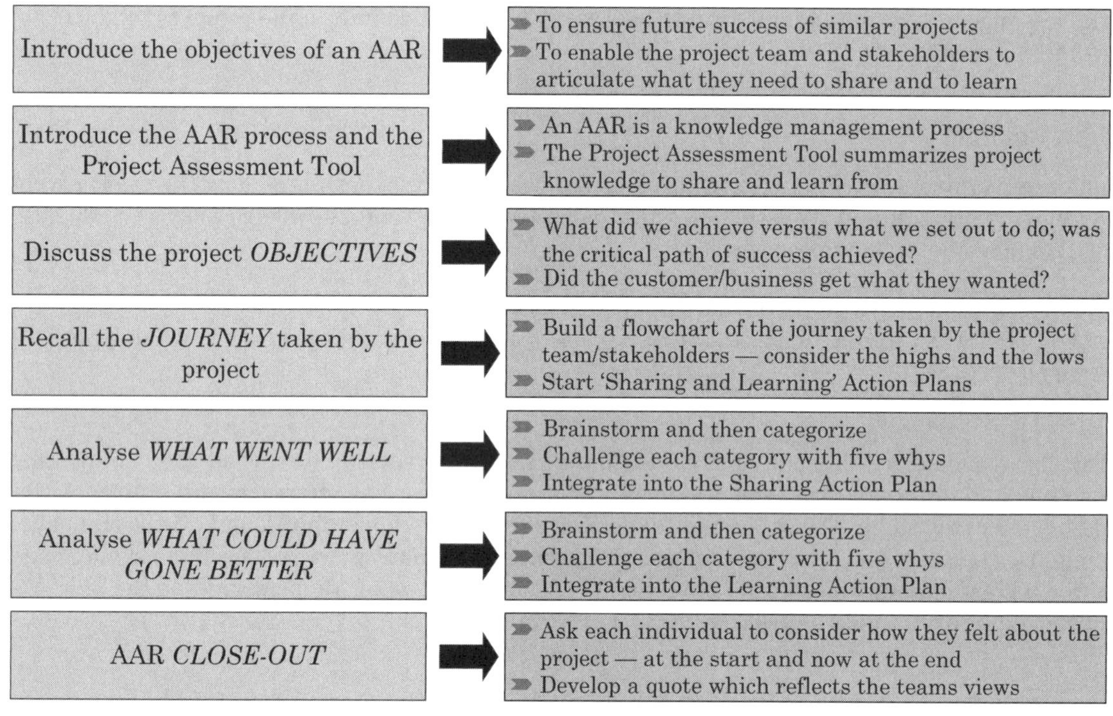

Figure 6-4 The after action review (AAR) process

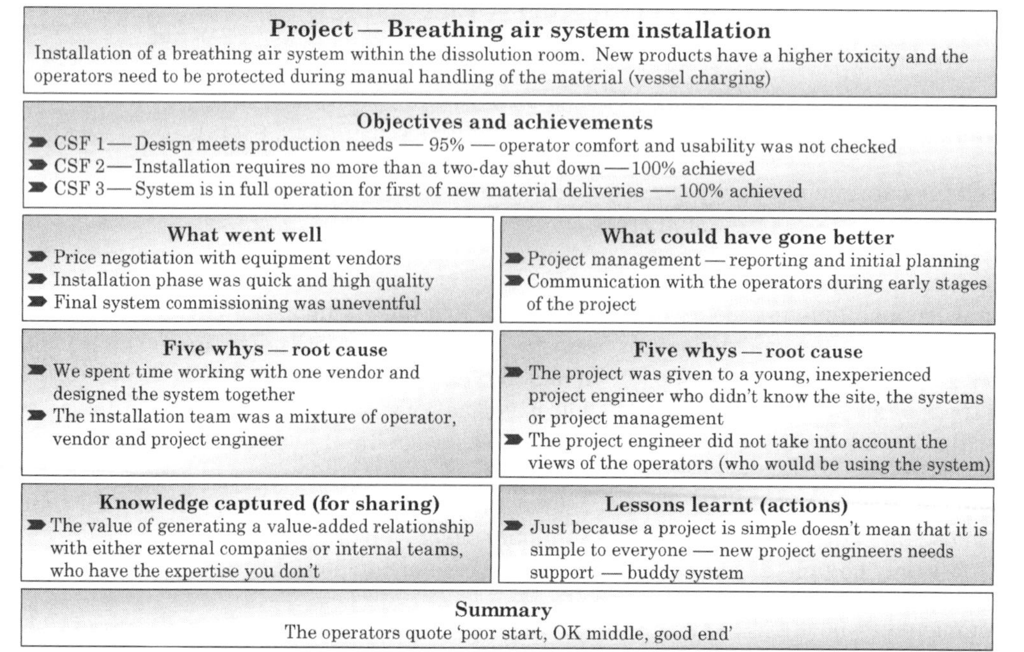

Figure 6-5 An after action review summary example — small project

Tool: Sustainability Checklist

The aim of this tool is to support the assessment of the sustainability of the changes associated with the project. The completion of this tool enables positive responses to the checks in the 'Benefits Realized?' Checklist introduced earlier in this chapter (see page 152).

The Sustainability Checklist introduces an agreed process for the sponsor and the customer which allows agreement on the sustainability of the change and also any subsequent actions required should the sustainability not be at an acceptable level.

The Sustainability Checklist explained

The checklist has four separate areas:

- Project description.
- Sustainability review information.
- Sustainability checks.
- Summary comments and next steps.

Table 6-7 shows the Sustainability Checklist with high-level guidance on how to complete each section. The following are additional, more detailed, notes to support tool completion.

Project vision

The vision of sustainability should be succinctly described here and relate to the vision developed as a part of the collation of the critical path of success for the project.

Sustainability review information

This section captures important sustainability review management information such as who will be involved and when the last review was held. This usually raises two queries:

- Who should be involved?
- How many reviews are necessary?

All sustainability reviews should be completed in partnership with the customer end user representative — the person or team who have responsibility for the maintenance of the assets or changes which resulted from the project. Depending on the project type, and the scope of the Project Manager's responsibilities, the Project Manager may be involved. The project sponsor would certainly be involved and in most cases would require support from the Project Manager.

Within a business change project the boundaries are clear — the Project Manager works with the sponsor and customer until the change has been sustained. For more traditional capital engineering projects the boundary is equally clear but completely different — the Project Manager formally hands over the completed asset at the end of the project leaving the sponsor and customer to realize the benefits and sustain any changes.

There are probably many examples in industry where the above are not true but this appears to be a general trend. This means that in many cases a Project Manager has to go beyond the stated role and support the sponsor. This can reap benefits in improved sustainability.

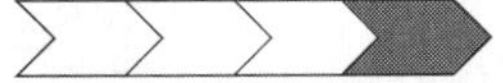

Table 6-7 The Sustainability Checklist explained

Project Management Toolkit — Sustainability Checklist

Project: *<insert project title>*	**Date:** *<insert date>*

Project vision

<Insert brief description of the vision of sustainability>

Sustainability review information

Previous sustainability review: *<insert date>* *<insert review number>*	**This sustainability review:** *<insert date>* *<insert review number>*
Project representative: *<insert name>*	**Customer representative:** *<insert name>*

Sustainability checks

Check number	Check	Target (sustained change)	Last review	This review
1	*<insert a check which will support the continued sustainability of the change which has been implemented>* *<note that sustainability checks do not track benefits>* *<insert as few or as many checks as necessary to assure all stakeholders that the change has been sustained>*	*<input a measurable target with appropriate units>*	*<insert measure or not applicable if this is the first review>*	*<note level of measure and insert any appropriate comments>*
2				
3				

Summary comments and next steps

<insert comments regarding the results of the sustainability check and any actions agreed with the customer>

Is the change completely sustained?	Yes/No *<delete as appropriate>*	**Date of next sustainability check**	*<insert date>*

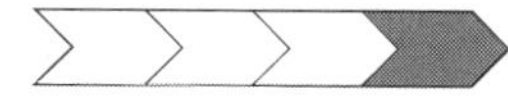

As for frequency of reviews — as many or as few as are needed to confirm that the change has been sustained.

In some dynamic technology based industries the final sustainability check can be the start of the next change.

On the basis that organizations are now managed by project this is becoming increasingly likely.

Sustainability checks

The most important yet hardest part of this tool is working out what needs to be measured in order to ensure the sustainability of the changes that have been made through the delivery of the project.

Sustainability checks are usually linked to the CSFs that have been delivered and reviewing them is a good start. The key is to define an activity or action that can be measured:

- How tidy a new production facility remains might be an indication of the efficiency of the production staff; however an untidy facility could be an indication of an under resourced facility.
- The change to a layout of a laboratory could be an indication that they wanted to go back to the 'old' ways of working; it could also mean that another change is required as the capacity has increased again.

A causal relationship needs to be identified between the things that can be seen, measured, checked and the original CSFs and the benefits being measured. Remember:

- Benefits metrics may become operational measures; sustainability checks do not.
- Sustainability checks do not track benefits.
- There is no limit to the total number of checks — the key is to have as few or as many as are necessary to assure all stakeholders that the change has been sustained.

Noting that sustainability checks may be less tangible and less measurable than the benefit metrics or CSFs, a 'target' still needs to be in mind — sustainability is achieved when:

- It is seen that the production area notice boards are 'live' and in daily use.
- Maintenance logs are reviewed and seen to tally with the preventative maintenance plan.
- Operational team behaviours are noted to be more about 'sharing is power' than 'knowledge is power'.

The actual nature of the check should hopefully become clearer in the case studies contained at the end of this chapter.

Summary comments and next steps

The final section of the sustainability review involves an agreed summary of the status of the sustainability — agreement between the project sponsor and the customer. They should then agree on whether a further review is needed. If not the customer is now accepting responsibility for the ongoing delivery of the business benefits and the sponsor's role is complete.

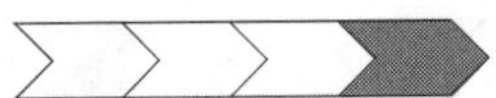

Using the Sustainability Checklist

Sustainability is a 'soft' disengagement process for the sponsor and, in some cases, for the Project Manager and is based on understanding the difference between a sustainability check and a benefit (Table 6-8).

A sustainability review can be a meeting, a tour of an area or a series of one-to-ones with a sample of end users. However, it has the most power when it involves entire teams of people affected by the change.

The very act of checking sustainability can actually have a positive impact on sustainability if the measures are appropriate — measures determine behaviours.

Table 6-8 Sustainability checks versus benefit metrics

Project	Example benefit metric	Example sustainability check
Engineering improvement	Plant reliability measured as up-time percentage	Preventative maintenance review versus procedures
New manufacturing facility	Cost to make product measured as £/Kg	Facility is operated and maintained in line with the design
Product launch	Market share as a percentage in that product group	Supply chain operates as designed
Organizational development	Reduced operating costs measured as £/site area	Procedural compliance

Case study J — if only the 'Benefits Realized?' Checklist had been around then!

This case study is based on a real example of a successful project that delivered no real business benefit. The 'Benefits Realized?' Checklist is used here to trouble-shoot — 'why were there no benefits?'

Operational changes

Situation

A site manager was concerned by the increasing energy costs at his site. His fears were confirmed when a global review of energy costs at each of the companies' sites showed his as one of the highest energy users. His was one of the oldest sites in the network.

The site manager knew that the main problem was likely to be related to the older, less efficient equipment at his site and so he instructed the engineering department to start a project to review all assets on site and to assess which older assets should be replaced.

The project

The engineering manager couldn't quite believe his luck — review all older assets and start to replace them! He had been fielding complaints about unreliable equipment and decreasing plant capacities over the last year and with a reduced capital plan he had not expected to be able to do anything more than some minor engineering maintenance.

He assigned his most senior Project Manager to the project and was personally involved in all major decisions.

The Project Manager pulled together a team of engineers from around the site and together they developed a robust scope linked to increasing the reliability of the highest priority production plant. New technology was reviewed, automation was considered, and best practice engineering solutions were all investigated to ensure that the problem was effectively and efficiently resolved.

The outcome

The project was well planned and well managed and finished ahead of schedule at only 3% above budget. The engineering manager considered it a complete success and reported as such to the site manager. He was hopeful that he could now move on to reviewing some of the smaller, lower priority production units, which were equally 'unreliable'.

Although the complaints from this production plant to the engineering department did cease there was no overall assessment on whether the plant reliability had increased. The plant continued to produce at the same rate, with the same lead-time being offered to customers. The site continued to meet its customer obligations — as it had before.

The site manager didn't see any decrease in his energy costs no matter which way he split the total cost (for example per employee, per Kg product made) yet he could see that the project to evaluate the old assets had been a success with investment in newer assets on one of the largest production units. He wanted to know why? Following a brief assessment of the situation he concluded that:

- There was never any baseline data collected regarding energy usage in the production unit nor equipment reliability as linked to the impact of production throughput or lead-times (Table 6-9).

Table 6-9 A completed 'Benefits Realized?' Checklist — case study J

***Project Management Toolkit* — 'Benefits Realized?' Checklist**	
Project: Operational changes	**Project Manager:** Confidential
Date: Confidential	**Page:** 1 of 2
Stage Three check	
Have there been any changes since Stage Three completion? (Note only the changes since the final Stage Three 'health check') Formal project reviews were conducted at a regular frequency during the project delivery. They showed that the project was well controlled. The project is now complete and this checklist is being completed as a part of a review into why no benefits appear to have been realized from this project	
Business benefits	
Has the business case changed since Stage One? (For example during planning and delivery, pre- or post-project approval) The formal business case stated that the capital expenditure would increase the reliability of the product supply chain whilst decreasing energy costs within the unit. The project was internally authorized by the site capital budget holder (Site Manager) The project category was 'security of supply' implying that customer supply would be compromized without the project	
Have all benefits been defined in terms of trackable metrics? (Why is the project being done?) No and recent issues have highlighted that there are issues with the maintenance budget and increased usage of other site resources such as utilities	
What is the customer feedback? (Feedback from all stakeholders in the business including the customer) The customer for the project was the production department — they are fairly happy with the upgraded equipment but haven't yet got it all working as they'd like	
Are the benefits being tracked? The only intangible benefit the engineering department has looked at is a decrease in calls from production to engineering The production department are not formally tracking anything other than normal operational metrics. These do not demonstrate any change, which could be directly attributed to the project	
Business change	
Is the business ready for the project? (If the project can only enable benefits delivery by changing the way people work — has this been delivered, for example training?) The production department are currently negotiating training courses with the equipment vendors	

(continued)

Table 6-9 (Continued)

Project Management Toolkit — 'Benefits Realized?' Checklist	
Project: Operational changes	**Project Manager:** Confidential
Date: Confidential	**Page:** 2 of 2
Scope definition	
Has the scope been delivered? Yes — the scope was agreed with the production team and was delivered exactly as agreed. Older equipment has been replaced with newer equipment. New technology and increased automation have been used	
Have the benefit enablers been delivered? (Will the project enable the benefits to be delivered now the project is complete?) No — the causal link between the various elements of the scope and the anticipated benefits was never made. Although from an engineering perspective some items of equipment may have increased in reliability, overall the production unit performance appears unaffected	
Stage Four decision	
Has the project been delivered? (Delivery of project critical success criteria) The project has delivered all it was designed to deliver within time and budget limits (minor overspend)	
Have the business benefits been delivered? (Why was the project done in the first place?) No — although it is suggested that from this point the energy usage of the production unit should be monitored. Additionally an assessment should be made regarding the overall reliability of the production unit rather than on an equipment-by-equipment basis	

Case study K — using the benefits realized tools to redevelop a project

This case study demonstrates what can happen when you use the tools on a project that appeared to have delivered benefits but where 6 months later the benefits had not been sustained. As a result additional work was completed to support the sustainability.

Engineering training records and standard operation procedures

Situation

In order to complete preparations for an external site audit the engineering department of a manufacturing organization employed an expert from a third-party company. The expert reported to the site engineering director and had his backing to complete any tasks necessary to ensure compliance within the engineering department prior to the audit.

One area, which the expert picked up on, was the training records held for the engineering department's standard operating procedures (SOPs). A review of the training requirements of the SOPs revealed that there was a shortfall in completed training and many training records were unavailable. The engineering department, like all other site departments, has a documentation controller who is responsible for ensuring that the training for SOPs is completed and records maintained.

The expert identified that as well as the shortfall in training needs there was also no SOP covering the requirements for training and maintenance of training records. This was identified as the root cause for the situation that had arisen, as the controller was not fully aware of his responsibilities. The expert therefore undertook a project with two tasks; the first to complete all required training and ensure that all records are available; the second to write the required training SOP and implement it.

The project

As the document controller would assume future responsibility for the system, the expert involved him in the completion of the two elements of work. The controller produced all of the required paperwork for the training records and compiled the resulting training records as they became available. The controller had input into the writing of the new training SOP that the expert produced.

The expert trained the document controller in the new SOP and generated the relevant training record. The expert ensured that all training was completed and chased up all other site departments and staff to ensure that they completed their training in time. The expert also ensured that any other new SOPs created by other members of the engineering department had all relevant training records completed.

Prior to the audit the project scope was successfully completed. The expert left the engineering department following successful completion of the audit — once the business benefits of the project had been delivered.

The outcome

A follow up review at monthly intervals during the engineering department compliance meeting has revealed that a new gap in training has opened up. There was a lack of training records for new SOPs created since the audit and for new workers who had joined to the company since the audit. A review of the training SOP showed that it did cover both of these scenarios and therefore the problem appeared to be with the document controller who had not completed the requirements of the SOP. The site manager asked for a review of the current situation and what should be done to bring the situation back under control. Any solution would need to be sustainable. Tables 6-10 and 6-11 outline the results of the review and the plan.

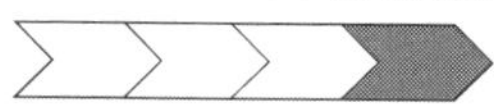

Table 6-10 A completed 'Benefits Realized?' Checklist — case study K

Project Management Toolkit — 'Benefits Realized?' Checklist	
Project: Engineering SOP training	**Project Manager:** Confidential
Date: Confidential	**Page:** 1 of 2
Stage Three check	
Have there been any changes since Stage Three completion? (Note only the changes since the final Stage Three 'health check') No — the requirements of the project remain the same: ➤ Complete all required training and records prior to the audit ➤ Prepare and implement new training SOP prior to the audit	
Business benefits	
Has the business case changed since Stage One? (For example during planning and delivery, pre- or post-project approval) No **Have all benefits been defined in terms of trackable metrics? (Why is the project being done?)** Complete — all required existing training prior to the audit is complete Track SOP training records requirements, ensure that a backlog of training requirements does not develop and that training is completed in a timely manner **What is the customer feedback? (Feedback from all stakeholders in the business including the customer)** All required training was completed. The engineering department passed the audit with flying colours, the Engineering Director received praise for his efforts **Are the benefits being tracked?** The training requirement versus records is reported monthly to the Engineering Director by the Document Controller	
Business change	
Is the business ready for the project? (If the project can only enable benefits delivery by changing the way people work — has this been delivered, for example training?) The engineering Document Controller was trained in the new SOP	
Scope definition	
Has the scope been delivered? The initial scope of the project has been delivered (the project department passed the audit with no findings). However the problem has occurred again and it is clear that the new training SOP has been insufficient in correcting the root cause of the problem. The Document Controller has been unable to ensure that the SOP is followed **Have the benefit enablers been delivered? (Will the project enable the benefits to be delivered now the project is complete?)** The perceived enablers were completed — SOP was prepared and training completed	

(continued)

Table 6-10 (Continued)

<table>
<tr><th colspan="2">Project Management Toolkit — 'Benefits Realized?' Checklist</th></tr>
<tr><td>Project: Engineering SOP training</td><td>Project Manager: Confidential</td></tr>
<tr><td>Date: Confidential</td><td>Page: 2 of 2</td></tr>
<tr><td colspan="2">Stage Four decision</td></tr>
<tr><td colspan="2">Has the project been delivered? (Delivery of project critical success criteria)
Yes.
Have the business benefits been delivered? (Why was the project done in the first place?)
Initially the benefits were delivered but this has not been sustained. If the site were to be audited now the engineering department would not pass. Discussion with the Document Controller and their opposite numbers in other departments has revealed the following issues. The Document Controllers had always been aware that the completion of training and maintenance of records was their responsibility. However they also suffered from two problems:
➤ Personnel generating new SOPs found the completion of the training difficult and laborious. There is a culture on site that people would not turn up to arranged training sessions and would not complete records. The view was therefore generally taken that it was better to at least have an SOP written and in place so that people could look at it when they needed to rather than have no SOP at all. The SOP writers were all senior to the document controller who was therefore powerless to prevent this activity
➤ The training issue also existed within other departments and had been an ongoing problem for some time. The document controllers were therefore often chasing each other to get training records and a blame culture had built up amongst them
The third party expert, with the backing of the Engineering Director, had the weight with which to get the job done and had managed to push the issue to completion. The status of the Document Controller role within the organization did not allow them to force the issue in this way. They also had the added problems of the site culture with regards to training generally and animosity amongst all the site Document Controllers who should have been working to help each other. A plan to sustain this change has been developed, see Sustainability Checklist (Table 6-11)</td></tr>
</table>

Table 6-11 A completed Sustainability Checklist – case study K

Project Management Toolkit — Sustainability Checklist				
Project: Engineering SOP training		**Date:** Confidential		
Project vision				
Training for engineering department SOPs should be up to date at all times. Records for completed training should be available on request as would occur during an audit				
Sustainability review information				
Previous sustainability review: Not applicable		**This sustainability review:** Not applicable		
Project representative: Engineering Director		**Customer representative:** Quality Director		
Sustainability checks				
Check number	**Check**	**Target (sustained change)**	**Last review**	**This review**
1	Ensure that anyone writing a new SOP is first trained in the training of SOPs	No new SOPs to be written by a person not trained in the training of SOPs	Not applicable	None, only Document Controller presently trained
2	SOPs will not be allowed to be considered 'live' until training has been completed and records produced	No new SOPs to be added to the SOP management system without completed training records	Not applicable	Not applicable
3	SOP writers to complete new training backlog within 1 month	Zero training record shortfall in 1 month other than new starters with company since today's date	Not applicable	Not applicable
4	Document Controller to issue training materials to all new starters within 1 month of arrival	New starters to be trained within 2 months of arrival	Not applicable	Not applicable
5	Personnel Manager requested to ensure that new starter information is issued to departmental managers in prompt and timely fashion	No new starters of which the engineering department is unaware of within 1 week of start	Not applicable	Not applicable

(continued)

Table 6-11 (Continued)

<table>
<tr><th colspan="5">Project Management Toolkit — Sustainability Checklist</th></tr>
<tr><td colspan="3">Project: Engineering SOP training</td><td colspan="2">Date: Confidential</td></tr>
<tr><th colspan="5">Sustainability checks</th></tr>
<tr><th>Check number</th><th>Check</th><th>Target (sustained change)</th><th>Last review</th><th>This review</th></tr>
<tr><td>6</td><td>Site Manager requested to issue a company wide communication concerning the completion of training for employees</td><td>Attendance at training sessions to be 90% or better</td><td>Not applicable</td><td>Not applicable</td></tr>
<tr><th colspan="5">Summary comments and next steps</th></tr>
<tr><td colspan="5">This was the first sustainability review that was completed due to the identified sustainability issues. This review proposed a plan around the above six sustainability checks</td></tr>
<tr><th>Is the change completely sustained?</th><td>~~Yes~~/No</td><th colspan="2">Date of next sustainability check</th><td>Within one calendar month</td></tr>
</table>

Case study L — using the benefits realized tools to review a project

This case study uses all of the Stage Four tools and is based on a real example in industry. The feedback at the end of the project was 'the benefits were delivered and sustainably!'

Overseas new facility build project

Situation

A pharmaceutical company had arranged a product licensing agreement with an overseas company for a bulk intermediate product. In return the overseas company would receive support in designing and building a new facility to make the intermediate product.

Although the overseas company had previously manufactured both sterile and non-sterile active pharmaceutical ingredients they had no experience of operating in compliance with US and European regulatory standards. All agreed that the biggest risk to the delivery of benefits from this venture would be the lack of an ongoing supply of compliant product.

The project

A project director with a background in pharmaceutical engineering projects and in managing regulatory strategy for new products was brought in to support the project. She liaised with local and European engineering firms, equipment vendors and sub-contractors.

Additionally she built a core team from the overseas production and quality assurance staff so that the culture of compliance was bought into.

The critical success factors and benefit metrics were defined.

The outcome

The project was successfully completed and after handover was able to produce the required three batches of product which could be tested and analysed so that appropriate data was available for the formal regulatory submission.

Due to the extent of the changes the production staff had to deal with — in going from a non-good manufacturing practice to a good manufacturing practice culture — the production manager agreed to quarterly visits by the Project Manager so that the sustainability of the changes could be monitored. This was in addition to the benefits tracking which was integrated into the production operational performance measures.

Initial reviews demonstrated that the change to different ways of working had been a harder transition than expected: good manufacturing practice issues increased, as did the level of batch rejects. The benefits tracking clearly demonstrated this lack of sustainability.

Action plans were put in place to provide additional training and auditing and the final two sustainability reviews demonstrated that this strategy had worked.

Table 6-12 A completed 'Benefits Realized?' Checklist — case study L

Project Management Toolkit — 'Benefits Realized?' Checklist	
Project: Overseas project	**Project Manager:** Confidential
Date: Confidential	**Page:** 1 of 2
Stage Three check	
Have there been any changes since Stage Three completion? (Note only the changes since the final Stage Three 'health check') None — this checklist was completed as a part of the final sustainability review (Table 6-13)	
Business benefits	
Has the business case changed since Stage One? (For example during planning and delivery, pre- or post-project approval) No — the benefit of having this facility overseas relates to the location of the licence for the intermediate to be manufactured there **Have all benefits been defined in terms of trackable metrics? (Why is the project being done?)** Yes — the benefit is having 24T/annum of intermediate available for the manufacture of the final drug product in the US or Europe at an agreed cost to manufacture **What is the customer feedback? (Feedback from all stakeholders in the business including the customer)** The Production Director and Site Director were pleased that the project proceeded according to plan and delivered the required benefits. Approval to manufacture was achieved and product was made and distributed to their licensing partner, who was equally happy with the situation **Are the benefits being tracked?** Yes — see Figure 6-6	
Business change	
Is the business ready for the project? (If the project can only enable benefits delivery by changing the way people work — has this been delivered, for example training?) Early on in the project it was identified that the development of a good manufacturing practice culture was necessary for the sustainable delivery of the benefits — ongoing compliant production. Throughout the project at each stage: design, procurement, construction, qualification and commissioning, the Project Team from the pharmaceutical company led by the Project Director would conduct training sessions for the production personnel The design of the plant in terms of layout and operation also built in good manufacturing practice compliance, for example the only way on to the plant was through the changing room — reminding all operators that they needed to change from their outdoor clothes before entering the plant	
Scope definition	
Has the scope been delivered? Yes — facility does produce an average of 90% of design capacity (24T/annum = approximately 2T/month) **Have the benefit enablers been delivered? (Will the project enable the benefits to be delivered now the project is complete?)** The project has been handed over in a compliant and fully operational state. Additionally all batch documentation has been developed jointly with the overseas production team	

(continued)

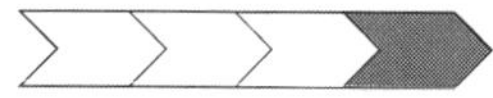

Table 6-12 (Continued)

Project Management Toolkit — 'Benefits Realized?' Checklist	
Project: Overseas project	**Project Manager:** Confidential
Date: Confidential	**Page:** 2 of 2
Stage Four decision	
Has the project been delivered? (Delivery of project critical success criteria) Yes — the facility has been handed over to production on time **Have the business benefits been delivered? (Why was the project done in the first place?)** Yes — production has commenced and the first three batches have been used to confirm that the product can be made reproducibly within pre-specified limits	

Table 6-13 A completed Sustainability Checklist — case study L

Project Management Toolkit — Sustainability Checklist

Project: Overseas project	**Date:** Confidential

Project vision

The project was the design and build of a new facility overseas, which would be capable of achieving approval to sell the pharmaceutical product. The facility must be capable of achieving and maintaining compliance to good manufacturing practice standards, local safety standards and 90% of the design of the plant throughout. Manufacturing costs are important as this links to the commercial deal between the two companies

Sustainability review information

Previous sustainability review: Review number 3 Date — confidential	**This sustainability review:** Review number 4 Date — confidential
Project representative: Project Director	**Customer representative:** Project Manager

Sustainability checks

Check number	Check	Target (sustained change)	Last review	This review
1	Number of good manufacturing practice deviations per operating shift (how operators comply with good manufacturing practice procedures as they work)	<5 minor 0 major	6 minor 0.5 major	5 minor 0.25 minor
2	Number of batch record deviations per batch (good documentation practices)	<5 minor 0 major	3 minor 0.25 major	4 minor
3	Training records	>90% accurate	100% accurate	100% accurate

Summary comments and next steps

The project was successfully handed over approximately 12 months ago and sustainability reviews have been held approximately every quarter

The benefits are being separately tracked (Figure 6-6)

Following the issues identified during the first two sustainability reviews two action plans were put in place: additional training and further auditing. These action plans have supported increased sustainability

Without the above changes sustainably in place it is clear that the benefits could not be sustainably delivered. When good manufacturing pratice deviations were at their highest — post handover — the cost of manufacture increased dramatically. Additionally due to the high level of reject batches at this time the overall plant throughput decreased (Figure 6-6)

The changes do now appear sustainable and a culture of good manufacturing practice compliance has developed. However I suggest an annual audit as a part of normal quality assurance procedures plus continued monitoring of the benefits. These would give an early indication of problems in this area

Is the change completely sustained?	Yes/~~No~~	**Date of next sustainability check**	Not applicable — benefits tracking will continue with annual audits of the above

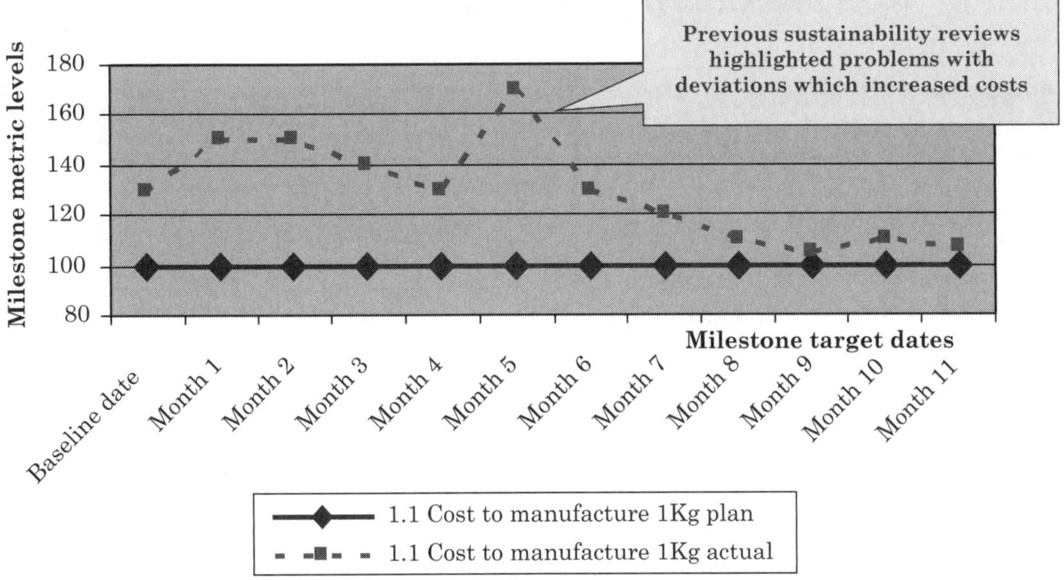

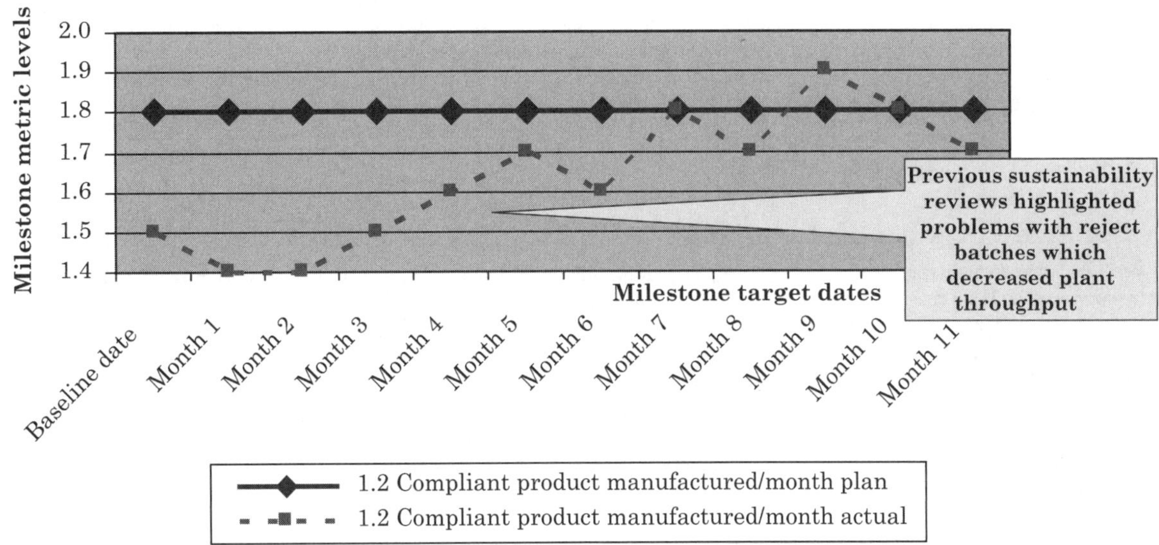

Figure 6-6 A completed Benefits Tracking Graph — case study L

Table 6-14 A completed Project Assessment Tool — case study L — an extract from the after action review conducted by the UK based team — the overall report is many pages and flip charts in length. All actions have subsequently been completed

***Project Management Toolkit* — Project Assessment Tool**		
Project: Overseas project	**Date:** Confidential	
Objectives review		
Objective/CSF	**Achievement**	**Action (sharing or learning)**
➤ Facility final cost £25M ➤ Schedule 18 months from concept to 1st Kg from the plant ➤ Enable regulatory approval to make and sell. ➤ Enable a production rate of 1.8T/month at a cost of $100/Kg	➤ Final cost £24.5M ➤ Schedule 17 months ➤ Regulatory Approval achieved within 12 months of handover ➤ Sustainable plant throughout of 1.8T/month at a cost of $110/Kg	➤ The value management activities to support design solutions ➤ The overall approach taken with the schedule — we were originally only accountable for mechanical handover but we scheduled to regulatory approval to support the overall project benefits realizations
Journey review		
The UK-based team who visited the overseas site on a regular basis described each phase as 'distinctly different' — they enjoyed the overseas travel experience and found the design phase and commissioning phases in particular to be 'different' — mainly due to the approach taken to integrate the business change issues into the project		
Highs	**Sharing**	**Learning**
➤ Watching the first Kg come out of the plant	➤ The sense of team achievement and the value in building a robust team	➤ Team building is value-add — you have to make sure it fits the team and organization culture
➤ Knowing that regulatory approval had been received	➤ Project Teams can benefit from looking at the bigger picture	➤ Understand what really matters to the customer
Lows	**Sharing**	**Learning**
➤ Seeing the clean room heating ventilation and air conditioning switched off each night as we left the site and having to 'make a stand'	➤ If you're passionate then people listen to you — there is no harm in showing you care	➤ Culture is hard to change and you just have to find ways to support it

(continued)

Table 6-14 (Continued)

Project Management Toolkit — Project Assessment Tool		
Project: Overseas project	**Date:** Confidential	
'What went well' review		
What went well?	**Why?**	**Sharing action**
➤ Design phase	➤ We had an integrated team from three countries and four companies — we learnt from each other — we had fun	➤ Share the team set-up and ways of working and how training in good manufacturing practice was integrated into the design process
➤ Commissioning and qualification	➤ We were a part of the production team. The Project Director had the final production say on good manufacturing practice issues	➤ Share the strategy for commissioning and qualification — the way that we effectively merged the project into the production team to deliver the result
'What could have gone better' review		
What could have gone better?	**Why?**	**Learning Action**
➤ Procurement	➤ We initially struggled with the overseas vendors that we thought couldn't provide the quality of equipment needed — they did	➤ Don't close your minds to opportunities — the overseas equipment was 25% cheaper and we had the time to sort out design issues
Project summary		
The Overseas Production Manager said 'the Project Director was fierce in protecting good manufacturing practice'		

Handy hints

Never be afraid to check if the benefits have really been delivered

It might not be your job but it will affect how people view the project you have delivered.

Support your sponsor throughout benefits delivery

The chances are that as soon as you have delivered the project scope you will be assigned a new project. However you did make a deal with your sponsor and so before you completely disengage from the project do check that the benefits realization plan is in place.

Never be afraid to keep a lower level involvement to ensure ultimate project success

It is too easy for Project Managers to move onto the next project, after all you are usually under pressure to do so, however there are key activities at the end of a project which support the organization, for example sustainability checking and after-project learning. If you want to achieve ultimate project success — sustainable business benefits in line with the organizational need — then you have to maintain some involvement with the key stakeholders.

Don't forget to let go and let the business reap the successes!

It is also equally easy for a Project Manager to remain in a project too long so that their role almost becomes a part of the 'way things work around here'. This is fatal! The Project Manager will eventually have to move on so to become indispensable to the ongoing success of the project is wrong. You need to handover to a customer representative and become dispensable!

Always look to expand your Stage Four toolkit

The Stage Four toolkit contains four tools out of many that are actually used in the realization of business benefits from a project. There are other tools that will help you — search them out and use them appropriately and pragmatically.

Further reading

Business benefits management is a crucial process within project management — no longer can we rely on just the delivery of our old friends cost, scope and time. These aspects are covered in more detail by *Project Benefits Management: Linking Projects to the Business* a further book in this Project Management Series (ISBN: 978-0-75068477-4).

Additionally the websites referenced on page xi can be used to source further 'best practice' material.

And finally . . .

- Ask 'have the benefits been realized' — if you don't know then the chances are they haven't!
- Remember — sustainability is the ultimate measure of project success.
- Be flexible — use and adapt the tools to achieve your goals.

7 Case Study One: the pharma facility project

This case study demonstrates use of the tools within a capital project, specifically the design and construction of a refurbished multi-purpose pharmaceutical manufacturing facility. It has been used for a number of reasons:

1. *Project type*

This is a capital engineering project. Such projects usually have very well-defined funding authorization and management procedures which are specific to an organization. Within this case study:

- The capital funding request procedure supported the development of a robust project scope with associated cost estimate.
- The financial management procedures restricted some of the project planning options, for example how any item of external expenditure was approved.

Organizational stage gate processes to manage capital funding can be an aid to good project management as they put 'hard' go/no go decisions in place. In the process used here, early stage gates match with the 'why' and 'how' decisions. This will support the project if the organizational business processes requirement for data match with the challenges laid out in the two associated stage checklists.

Figure 7-1 shows a typical stage gate process for a Pharma Capital Project. Alongside the capital funding approval stage gates there are also 'critical quality compliance stage gates'. One main difference from a normal capital project is the requirement to record and document criteria specified as having an impact on product quality throughout the project lifecycle. This ensures that these requirements are included and operate as planned within the finished plant.

Normally a capital project would not receive full funding until some predetermined level of design had been completed. This gives an assurance that the funds requested would be sufficient to deliver the project. The level of rigour in business case development depends on the link between capital funding approval and the need for the asset associated within the capital.

Typically, following capital approval, an organization will have strict guidelines on the management of the approved funds. Although supporting overall financial control and management of contingency this can restrict appropriate procurement practices, for example capital equipment sourcing strategies, corporate procurement compliance rules, etc. This is not atypical and the impact on project management will be discussed as the case study progresses, for example how inflexible procurement practices can conflict with project priorities.

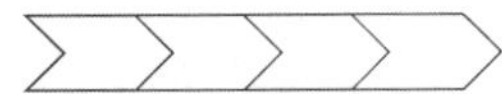

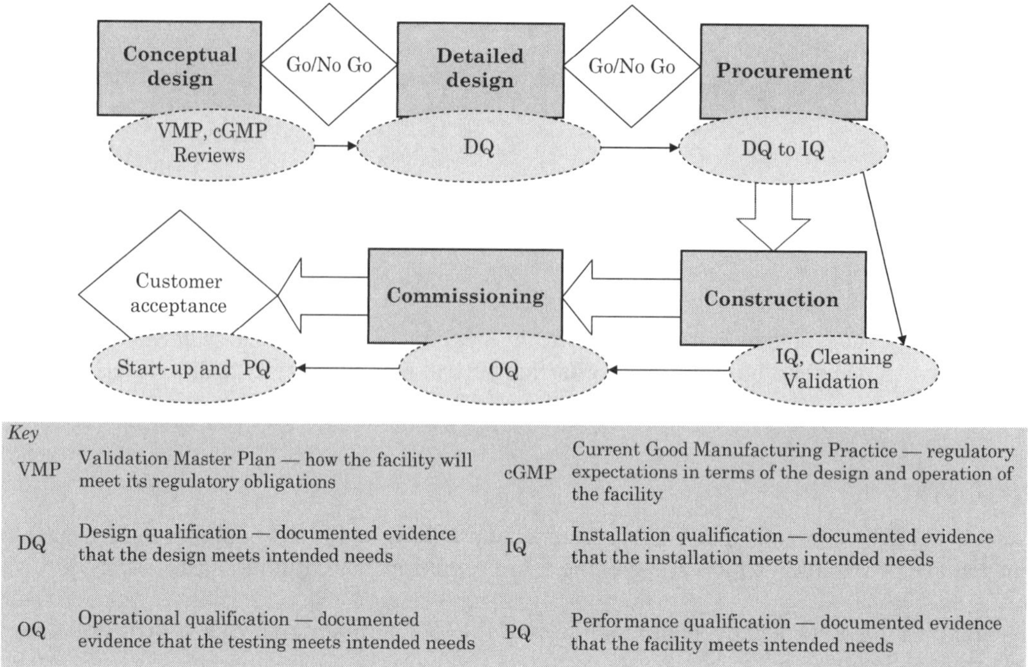

Figure 7-1 Pharma project stage gate process

2. Pharmaceutical environment

A facility project of this type (refurbishment) being executed within a highly regulated and competitive environment has some very specific challenges:

- The criticality of understanding the cGMP drivers (current good manufacturing practice) so that all 'risk to patient' is eliminated. This is achieved through integration into all project activities to ensure that compliance risk is effectively and proactively managed.
    - The goal of cGMP is to protect the product from contamination and ensure *consistent* manufacture to a predetermined specification and within specific critical operating parameters.
- The nature of regulatory compliance through the development of facilities that can be qualified in a way that is sustainable through the life of the facility. Qualification is the way that equipment, facilities and systems are proved capable of consistently performing as designed within their specified limits.
- During product development a 'Design Space' is specified. This defines the processing limits (for example temperature, pressure) within which the clinical trials material will have been manufactured. This information will be included within the New Drug Application (NDA) which is issued to regulators as part of their process for licensing the plant to manufacture when it is completed. Therefore the facility must be designed and constructed so that the product can be made in the same way that the clinical material was made.
- The competitive nature of the pharma market that wants drugs which cost less to manufacture and distribute as well as new drugs on the market quicker.
- The organizational drivers to maximize asset utilization in order to deliver increased return for the company can lead to an inappropriate focus within a project. Within a facility refurbishment

project there will be a requirement to shutdown the facility in order to complete the installation and testing of new equipment, etc. Often the cost of the shutdown is passed onto the new product, although this will depend on an organization's accounting policies. Whatever the specific driver the outcome is the same — pressure on the Project Manager to minimize the shutdown.

Therefore it is useful to demonstrate how the project tools in this book were used alongside other industry-specific tools to support completion of a pharma facility project. Examples of the generic tools which support qualification and cGMP compliance are:

- Document matrices — the management and control of documentation is fundamental to qualification, to proving consistent performance against specification.
- Risk assessments — applied specifically to the quality aspects of the project to mitigate against any potential 'risk to patient'.
- Tools that support the management of external suppliers — the use of pre-qualification, qualification and contract plans supports the selection and management of the appropriate skills required to achieve a qualified facility.
- Project delivery plans — structured planning is required to plan for successful qualification.

A pharma project does require technical expertise, but like any project this cannot be held solely within the Project Manager; an appropriate team needs to be pulled together. So although a Pharma Project Manager does not need to know all the answers, he needs to ask the right questions and then act on the answers appropriately.

3. Project outcome

Although this project was deemed to be a success it did not proceed smoothly and required many 'interventions'. The nature of these involved independent advice from an internal project management expert.

The root cause of the detailed issues is discussed in the individual stage 'Lessons learnt' reviews. However, the overall root causes of the sum of all the issues which occurred during the project life were:

- Assigning an inexperienced Project Manager to a large and complex capital project.
- Allowing corporate procurement strategies to drive the overall delivery strategy.
- Unclear and haphazard sponsorship — early in the project lifecycle from both sides of the sponsor contract.
- Corporate measures of operational success driving behaviours which distracted the Project Team from higher-priority goals.
- Inability to define what was really critical to the project at an early enough stage.
- Unclear definition of the business benefits.
- Reaction to inaccurate early feasibility work that defined a cost and schedule target range, yet was used as an absolute target.

4. Actual tool use

The toolkit was used in a number of ways on this project:

- By the internal project management expert to assess the 'health' of the project at various stages and to support development of appropriate intervention strategies.
- By the Project Manager to proactively support effective project delivery — noting that the Project Manager was going through training during the delivery of this project.

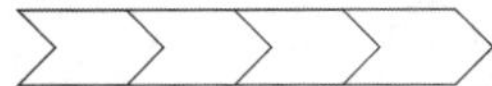

Situation

A pharmaceutical company is currently conducting Phase III clinical trials for a new product, XP515. This product will have a significant impact on the lives of patients currently using a competitor drug; however, the company needs to move fast if it is to fully leverage this. It is known that the competitor is entering Phase III trials for its own improved drug. This is a new therapeutic area for the company and therefore a new market.

A launch timetable has been developed, and forecast sales have indicated a level beyond that previously anticipated. Although there is sufficient secondary capacity at the chosen site location (formulation and packaging), there is insufficient primary capacity (the chemical synthesis of the bulk active pharmaceutical ingredient — API). Therefore a new site of primary manufacture is needed. There are five options:

1. Expand the multipurpose clinical manufacturing facility at the current site.
2. Use one of the current API multipurpose plants located at another site.
3. Build a new facility at a current site location.
4. Revamp a current bulk manufacturing facility at another site which has spare capacity.
5. Use an external contract manufacturer.

The evaluation of the five options was based on three criteria: business risk, cost, and time (Table 7-1).

Table 7-1 XP515 manufacturing strategy

Manufacturing option	Business risk	Cost	Time	Other comments
1. Clinical facility	High	Low	Low	This asset cannot be dedicated to XP515 due to other product development needs Unlikely to be able to be expanded beyond initial launch needs, so would still need another facility in 6–12 months
2. Existing API facility	High	Low	Low	All the current API-manufacturing facilities are operating at the maximum capacity No other product campaign can be delayed to substitute for XP515
3. New facility	Medium	High	High	Although the forecasts are good, the business could not rely on one product to fill a new facility The capital costs would be high and the likely lead-time would delay XP515 launch
4. Revamp a bulk facility	Medium	Medium	Medium	There are two possible sites (UK and Spain), although neither has experience of final-stage API manufacture The facilities have the spare capacity to handle XP515 as well as meet current commitments
5. Contract facility	High	Medium	Low	Some intermediates and most excipients are supplied in this way; however, it is not currently used for API manufacture due to the high supply chain risk that this would introduce

The fourth option is seen as the most viable for the company considering all aspects: maintaining a competitive cost of goods, achieving the launch timetable, achieving regulatory compliance and aligning with longer-term business goals. Of the two sites being considered, the UK has been selected on the basis that early feasibility studies have shown:

- The facility could be revamped at a cost of $10 million (±30%) as opposed to $14 million in Spain.
- The UK facility would be ready to manufacture the new drug 3 months earlier than the Spanish site due to the difference in the scale of the refurbishments.

The UK Facility 2B is an existing multi-product primary pharmaceutical manufacturing facility that has never manufactured final-stage APIs. It usually manufactures early stage intermediates with minimal criticality to the final drug substance or excipients. The company has often thought about assessing the feasibility of using the facility in a more flexible way; so this is seen as an opportunity to refurbish for potential future API use, including the manufacture of XP515.

The corporate engineering group which oversees all major projects is extremely busy at present and has allocated a relatively inexperienced Project Manager (Fred Jones) to the project as it is seen as relatively straightforward (revamp a facility to meet cGMP expectations). Additionally Fred has worked on the site before and has a good working relationship with the site Production Director (Bob Smith).

Fred will need to follow capital funding authorization procedures and then, following approval, local and corporate finance management procedures.

The following sections track the project through each stage of its lifecycle as linked to the specific roadmap for this type of project (Table 7-2 and Figure 7-1):

Table 7-2 Project stages vs the pharma facility project lifecycle

Project stage	Pharma facility project lifecycle	Comments
Stage One	Capital plan review	Confirmation that the business needs the project — usually a capital allowance is made based on early feasibility work
	Conceptual design	Early design work to achieve a ±20% cost and schedule accuracy against a draft scope
End of Stage One and Stage Two	Front-end design	Sufficient level of design to define the scope and the associated ±10% cost and schedule which then becomes the basis for the business case. This usually requires a confirmed method of project delivery that links scope, cost and time
Stage Three	Detailed design	Completion of the design so that all items within it can be procured and installed
		Requires some design data from external suppliers in order to be complete; so this stage and procurement are iterative
	Procurement	All items in the design are procured according to the defined contract strategy (links to the delivery plan)
	Construction	All items in the design are constructed and/or installed according to the defined construction strategy (links to the delivery plan)
		This stage links to any issues with site or facility access (whether new or revamp projects, greenfield or brownfield)
	Commissioning and qualification (C & Q)	All items in the design are tested according to the defined C & Q strategy (links to the delivery plan and the Validation Master Plan)
Stage Four	Handover	The facility is handed over to the team that will operate it and actually produce the product for which it has been designed
	Closure	Review of the project to determine if successful and final check that the project has been fully completed (financial closure, documentation complete and handed over, etc.)

Stage One — Business Case Development

In order to gain capital funding approval the project went through a series of early design phases. This allowed the development of an accurate scope, based on the initial site location decision estimates, an associated accurate capital cost estimate and overall project schedule. The conceptual design work was based on the last version of facility 'as builts' (approximately 5 years since the last major project in the facility) and determined the following project parameters:

- a capital budget of $12 million (±20%)
- project complete within 18 months
- a 6 month shutdown

The work was completed by an engineering design contractor selected by the corporate engineering group from their list of pre-approved engineering companies and in close liaison with the Project Manager Fred Jones.

One area of concern that has already arisen has been the Production Director Bob Smith's comments on the minimal time he is willing to allow the 2B plant to shutdown — he is concerned with his plant utilization's Key Performance Indicator (KPI). Fred knows that it is critical that Bob understands the cGMP drivers on this project and the need to revamp key areas of the facility where the API will be exposed to the environment. He is working with Bob to see how the plant utilization targets for next year can be lowered. Bob has agreed to a partial 4 month shutdown with some areas still being operational (one train of the equipment-making excipients).

Following completion of the estimate which was above the original amount of $10 million given at the site selection stage, Fred determined that it would be useful to revisit the business case. This would ensure a smooth decision-making process as the project moves forward for approval prior to detailed design. He therefore completed the 'Why?' Checklist (Table 7-3) which in turn prompted the completion of other available tools including:

- Simple Benefits Hierarchy (Figure 7-2).
- Benefits Specification Table (Table 7-4).
- Business Case Tool (Table 7-5).

Table 7-3 The 'Why?' Checklist

***Project Management Toolkit* — The 'Why?' Checklist**	
Project: XP515 Upgrade	**Project Manager:** Fred Jones
Date: End Month 3	**Page:** 1 of 2
Sponsorship	
Who is the sponsor? (The person who is accountable for the delivery of the business benefits) My link has been through the Production Manager, Bob Smith, and at this stage it is not clear who is taking the sponsorship role. I have asked the question and assume that the XP515 Product Manager (customer) will take on this accountability **Has the sponsor developed an external communication plan? (How the sponsor will communicate with all stakeholders in the business)** There is no clear communication outside of the engineering group other than with Bob Smith	

(continued)

Table 7-3 (Continued)

<table>
<tr><th colspan="2">Project Management Toolkit — The 'Why?' Checklist</th></tr>
<tr><td>Project: XP515 Upgrade</td><td>Project Manager: Fred Jones</td></tr>
<tr><td>Date: End Month 3</td><td>Page: 2 of 2</td></tr>
<tr><td colspan="2">Business benefits</td></tr>
<tr><td colspan="2">Has a business case been developed?
Yes — I have made a number of assumptions relating to the importance of this product to the company and would like to discuss with the sponsor
Have all benefits been identified? (Why is the project being done?)
Yes — this is clearly about selling a new product and doing so from a modified facility. The benefits of using this facility appear to be cost related and there appears to be a fixed launch date
Who is the customer? (Identify all stakeholders in the business including the customer)
Assume the XP515 Product Manager is both the customer and the sponsor
How will benefits be tracked? (Have they been adequately defined?)
Unclear at present as have not had direct contact with the customer. ACTION — develop a Benefits Specification Table (Table 7-4)</td></tr>
<tr><td colspan="2">Business change</td></tr>
<tr><td colspan="2">Will this project change the way people do business? (Will people need to work differently?)
Yes — the increased cGMP operational requirements will change the way that the facility runs on a day-to-day basis
Is the business ready for this project? (Are training needs identified or other organizational changes needed?)
Not yet — this needs to be integrated into future communications with Bob Smith and also into the project plan</td></tr>
<tr><td colspan="2">Scope definition</td></tr>
<tr><td colspan="2">Has the scope been defined? (What level of feasibility work has been done?)
Only in outline — early feasibility work identified that this facility could be revamped at a cost of $10 million and that this would deliver a facility that was capable of achieving regulatory approval for the manufacture of XP515. However conceptual design work has further clarified the scope and associated capital cost ($12 million)
Have the benefit enablers been defined? (Are you sure that the project will enable the benefits to be delivered when the project is complete?)
To achieve XP515 approval in this facility the internal layout needs a complete review with many additional segregated areas needed including one clean room (for final product kegging)
Have all alternatives been investigated? (Which may include not needing the project)
Bob Smith seemed to indicate that his facility had been in competition with the one in Spain for this product and that the feasibility work has highlighted that the costs of doing the project in the UK were lower. Current existing facilities which have the appropriate level of cGMP have no capacity for this new product
Have the project success criteria been defined and prioritized?
Yes — achieve the launch timetable, ensuring that the facility is qualified so that the regulatory submission can be completed</td></tr>
<tr><td colspan="2">Stage One decision</td></tr>
<tr><td colspan="2">Should the project be progressed further? (Is the business case robust enough for detailed planning to commence?)
Yes — although some outstanding questions remain:
➤ Who is the sponsor?
➤ When can the construction and commissioning of the revamped facility be done so that it meets both asset and product goals (linked to the agreed facility shutdown — date and duration)?
➤ Can the Project Team have access to the feasibility work which would include a facility impact assessment against cGMP needs for the product?</td></tr>
</table>

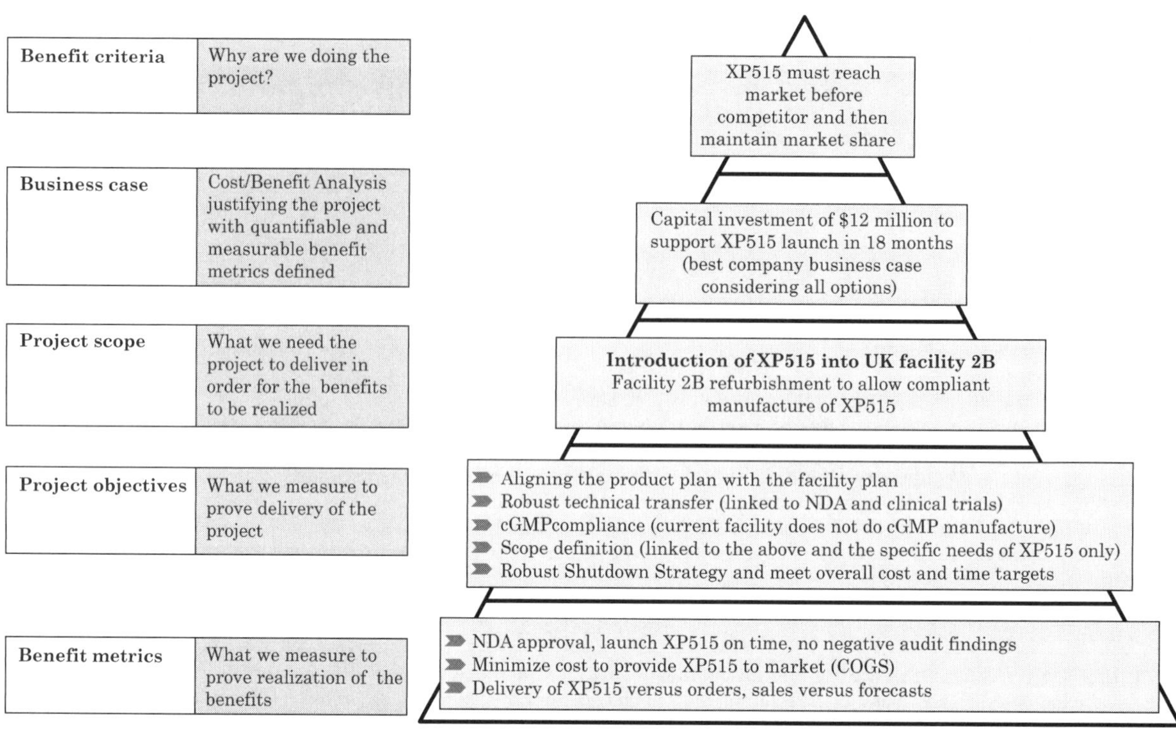

Figure 7-2 A Simple Benefits Hierarchy

These tools were completed during meetings between Fred and Bob with support from the original project documentation supplied to Fred when he commenced work on the project.

Table 7-4 Benefits Specification Table

Project Management Toolkit — Benefits Specification Table

Project: XP515 Upgrade			**Date:** End Month 3		
Potential benefit	**Benefit metric**	**Benefit metric baseline**	**Accountability**	**Benefit metric target**	**Area of activity**
What the project will enable the business to deliver	*Characteristic to be measured*	*Current level of performance*	*Accountable for delivery of the benefit to target*	*Required performance to achieve overall benefits*	*The project scope that will enable this benefit to be delivered*
XP515 meets and sustains market share requirements	Design, procurement and construction schedule completion dates	Project only recently started	Fred Jones	Achieve all major milestone dates	Facility project
	Facility is approved by external regulators to manufacture with no non-conformances or additional requirements prior to start up	NDA submission has been made by company technical operations department	TBC	No non-conformances or additional requirements	Detailed design Plant qualification
	Production capacity	Target for capacity set by process requirements and design	Bob Smith	Production capacity to be better than 75% of design during and after process qualification batches	Must be achieved following project handover in order for the project goals to be achieved
				Production capacity to achieve 95% of design within 3 months of start up	Must be achieved following project handover in order for the project goals to be achieved
	Manufacturing cost	Cost data for existing products based on staffing, site overheads, etc.	Bob Smith	Production costs and internal company charge per kg to meet target set within submitted site selection data	Must be achieved following project handover in order for the project goals to be achieved

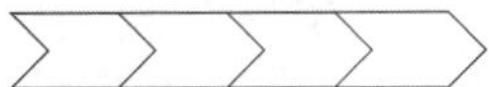

Table 7-5 The Business Case Tool

<table>
<tr><th colspan="4">Project Management Toolkit — Business Case Tool</th></tr>
<tr><td colspan="2">Project: XP515 Upgrade</td><td colspan="2">Date: End Month 3</td></tr>
<tr><td>Business case developed by</td><td>Fred Jones (Project Manager)</td><td>Date</td><td>End Month 3</td></tr>
<tr><td>Project reference Number</td><td>N/A</td><td>Business area</td><td>Corporate Technical Services</td></tr>
<tr><td>Project Manager</td><td>Fred Jones</td><td>Project sponsor</td><td>Currently Bob Smith (This may require development as the project progresses)</td></tr>
<tr><td>Business background</td><td colspan="3">Following Phase III clinical trials product XP515 has been identified as offering significant benefits over existing drugs within an existing target market. Manufacture and launch of XP515 before competitors will allow the company to achieve a significant market share and advantage. Based on conceptual information several opportunities for manufacture of the bulk API have been considered. Plant 2B at the UK site has been selected as the best option for meeting time and cost requirements. On the basis of the existing market and performance of the new drug a robust sales plan has been generated which has been used to define overall plant design capacity</td></tr>
<tr><td>Project description</td><td colspan="3">The project must modify an existing bulk manufacturing facility which presently manufactures intermediates and excipients. Whilst the plant contains 75% of the required equipment and service infrastructure several new items of processing and filling equipment are required. Improvement is also needed to containment and cGMP standards in areas where the product will be exposed to the plant atmosphere for manual vessel additions and product offloading</td></tr>
<tr><td>Delivery analysis</td><td colspan="3">Corporate engineering will support this project, providing project management, planning and cost control expertise
An external Engineering Contractor will provide engineering design services as these are unavailable from the corporate engineering group at this time and the site engineering group does not have sufficient resource
Automation and control design, procurement and software development will be completed by an in-house site-based department who have included the requirements for this project within its forward plan for the next year
Conceptual design work has indicated a project cost of $12 million (±20%) This cost is above initial estimates of $10 million for development of the UK 2B plant site
Operator orientation and training will be required. This has not been allowed for within the capital budget and must be provided by the site. A budget has yet to be determined and agreed
There is one other ongoing project in this area to replace existing chilled water process and circulation plant. This project has minimal shutdown requirements, is well progressed and is scheduled for completion 6 months ahead of the XP515 project</td></tr>
<tr><td>Business change analysis</td><td colspan="3">This project will allow the business to gain entry into an existing target market where it currently has no market share</td></tr>
</table>

(continued)

Table 7-5 (Continued)

<table>
<tr><th colspan="4">Project Management Toolkit — Business Case Tool</th></tr>
<tr><td colspan="2">Project: XP515 Upgrade</td><td colspan="2">Date: End Month 3</td></tr>
<tr><td>Value-add analysis</td><td colspan="3">Achieve XP515 launch date. This requires:
➤ Completion of facility modifications to schedule
➤ Approval of the facility and process from external regulators with no non-conformances or additional actions
Meet production capacity requirements:
➤ Plant to manufacture at least 75% of design target during pre-production trials and post-completion
➤ Plant to manufacture at least 95% of design targets within 3 months of start up
➤ Product change over and clean downs to proceed as scheduled
➤ 'Changes in other plant 2B product requirements to have no impact on XP515 bulk manufacture</td></tr>
<tr><td>Impact of not doing the project</td><td colspan="3">The impact of not completing the project will be to miss an opportunity to gain market share in a new target market and an advantage over the competition in this market</td></tr>
<tr><td>Project approved (Value add or not?)</td><td>Yes</td><td>Name of Approver and Date</td><td>UK Site Manager End Month 3</td></tr>
</table>

End of Stage One situation

- Including contractual arrangements to get the engineering contractor in place it has taken 3 months to get to stage gate approval for detailed engineering. The project now has in place:
  - A revised capital cost of $12 million, with a sound basis, which has been inserted into the site capital plan.
  - A high-level project plan, which meets the launch requirements, developed in conjunction with the Engineering Contractor and Production Director. Overall time line is 15 months from now to project completion.
- There is a business expectation that Facility 2B will continue to meet its asset utilization targets.
- The business drivers have been clearly articulated:
  - Drug manufacture and launch to meet sales requirement.
  - Gain regulatory approval for the manufacture of the drug from this revamped facility.
  - Achieve specified design capacity for the plant within 3 months of start-up.
- The conceptual design work by the engineering contractor has cost approximately $35 000.

Lessons learnt

A brief after-action review highlighted the following key issues:

- There is unclear sponsorship and the Project Manager hasn't got any form of action plan to deal with this.
- The requirements for shutdown have not been fully determined or agreed and the present situation seems untenable as the needs of neither side are likely to be achieved. The project requires too long a shutdown, which under present conditions the production department has not agreed to. The project will need to ensure that the requirements and benefits are fully understood, so that we can get agreement which satisfies all drivers — product and asset.
- An impact assessment has not been completed. This would give a better understanding of the business changes as related to the cGMP changes. The upgrade of the facility will require a change in attitude and capabilities for operating staff as well as possibly impacting quality testing and quality assurance staff.

Stage Two — Project Delivery Planning

Following project approval, a small team was formed to plan how this facility would be designed and then constructed so that the benefits could be realized as soon as possible.

Pre-sanction funding of approximately $100 000 was approved from Stage One and front-end design work was conducted with the External Engineering Contractor. The purpose of this additional level of design is to determine the final project sanction estimate and to complete design of any long lead single-source plant items required for the project. The types of engineering solutions identified were:

- Further physical segregation of the facility to support cGMP needs in terms of people and material flows.
- Requirement for at least one 'clean room' (an area with a controlled environment suitable for the exposure of the final API for product loading).
- The need for increased control of reactors due to the criticality of temperature within the new process — new reactor heat transfer systems and temperature control loops designed.

The Project Manager, Fred Jones, developed the project plan in liaison with the Engineering Contractor. Issues which came up during this time were:

- The project will need all existing documentation. Bob Smith (Production Director) will need to 'open doors' for the team rather than provide this himself.
- Apart from engineering specifications we need to review the materials which have been previously manufactured in the facility to assess the requirements for de-contamination.
- We need to understand utility availability as we currently have not considered this to be in scope. It will only come into the project if capacity or quality requirements drive it that way.
- It is likely that the project will need to get operations, technical services, maintenance and Quality Assurance lab personnel into the design team and then later into the commissioning team — we need to start this negotiation now. We need the names of potential team members and when they can be released.
- We need an early decision on *when* the facility can be shutdown and handed over, then we can incorporate this into the plan and build in robust schedule buffers.
- The engineering company does not have all the details required for design of the revised facility linked to the original clinical trial data.

It appears that Bob doesn't understand the implications of what we are doing to the facility and the increased cGMP operational requirements. Bob needs to use the shutdown time as an opportunity for operator training and also to allow time for the development of new operational and maintenance procedures.

A standard Project Team (Figure 7-3) has been put together comprising:

- External engineering team: engineers and designers.
- Internal automation and control group for procurement and software development.
- Corporate project engineering and finance support.
- Liaison to the existing facility and to QA support.
- Production Manager.
- Site technical services including process chemists.

During capital approval, planning and design phases the Project Manager has arranged to meet with Bob Smith every 2 weeks and to send him a copy of the high-level project report (monthly until construction and then weekly).

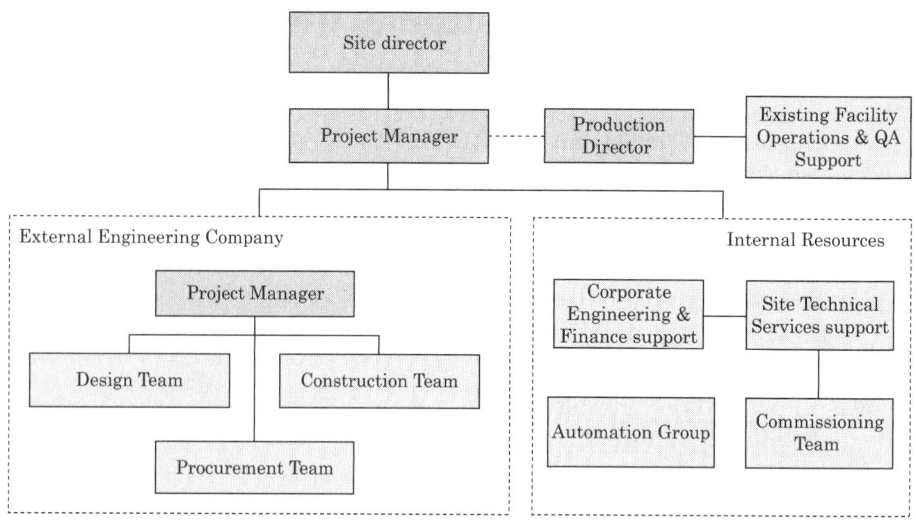

Figure 7-3 Project organization

The first task for this Project Team was to develop the path of success and the table of CSFs.

The path of success (Figure 7-4) was a relatively new concept for Fred, and initially he struggled to understand the 'big picture' as he felt most of it was outside of his engineering project remit. However the final 'test' for the CSFs proved that without each one being successfully achieved, it was unlikely that the project vision could be achieved.

CSF 1 **STRATEGIC ALIGNMENT** Ensuring that the product and facility strategy align with each other and with the organizational goals	**CSF 2** **ROBUST TECHNICAL TRANSFER** Robust transfer of technical process so that XP515 can be manufactured as per the clinical trial material and as defined in the NDA	**CSF 3** **FACILITY QUALIFICATION and cGMP COMPLIANCE** Ensure that all aspects of Facility 2B are cGMP compliant, qualified appropriately and can sustain that qualification
CSF 4 **APPROPRIATE SCOPE DEFINITION** Scope definition (linked to the above and the specific needs of XP515 only)	**CSF 5** **ALIGNED SHUTDOWN STRATEGY** The development and delivery of a robust shutdown strategy which meets the business needs	**CSF 6** **BUSINESS CASE DELIVERY** The project must meet the stated business case in terms of the capital cost approval and the schedule milestones

PROJECT VISION
Facility 2B refurbishment to allow compliant manufacture of XP515 to meet launch and post-launch targets

Figure 7-4 Path of success

In converting the path of CSFs into a Table of CSFs (Table 7-6) it became clear that there were gaps which needed to be filled.

Table 7-6 Table of Critical Success Factors

***Project Management Toolkit* — Table of Critical Success Factors**				
Project: XP515 Upgrade		**Date:** Start Month 4		
Critical path of success				
Plant 2B is to be upgraded for the manufacture of product XP515. Plant must be fully qualified and operational within 18 months in order to meet sales objectives for XP515 within a new market area				
Critical success factor definition				
Scope area (CSF Level 1)	**Objective tracking metric (CSF Level 2)**	**Critical milestone (CSF Level 3)**	**Accountable for CSF Level 3 delivery**	**Priority (within scope area)**
Strategic alignment	Overall 2B plant capacity	Overall design capacity will allow XP515 production at required quantities	Production Director	High
	Overall 2B plant product mix	Forecast product mix will not impact XP515 production or require onerous clean down for change over	Production Director	High
	2B utilities capability	2B utilities capacity to meet design requirements within schedule requirements	Production Director	Medium
Robust technical transfer	Design must meet requirements of NDA	Review at design qualification, no non-conformances	Project Manager	High
	Input of clinical trial data	Review at design qualification, no non-conformances	Project Manager	High
Facility qualification and cGMP compliance	Installation phase cGMP audits	All audit actions closed out at completion of installation	Project Manager	Medium
	Develop all appropriate documentation for qualification	Documentation is approved for execution in line with schedule requirements	Qualification Team Leader (Site Technical Services)	Medium
	Obtain all equipment documentation from contractors and vendors	Vendors supply of all documentation, material certificates, manuals, etc. will be included as part of final payment release	Document Controller (Engineering Contractor)	Medium
	Complete installation, operational qualification	Qualification complete within schedule including close out of all outstanding issues	Qualification Team Leader (Engineering Contractor)	Medium

(continued)

Table 7-6 (Continued)

***Project Management Toolkit* — Table of Critical Success Factors**				
Project: XP515 Upgrade		**Date:** Start Month 4		
Critical success factor definition				
Scope area (CSF Level 1)	**Objective tracking metric (CSF Level 2)**	**Critical milestone (CSF Level 3)**	**Accountable for CSF Level 3 delivery**	**Priority (within scope area)**
	External regulator inspection	No non-conformances or additional actions	Project Manager	High
	Process qualification	Operator training completed within schedule requirement	Production Director	Medium
		In process and final testing quality assurance staff trained	Site Manager	Medium
		Produce 3 concurrent batches with process in control: no non-conformances; no more than 1 minor deviation from operating norm per batch	Production Director	Medium
Appropriate scope definition	Control of scope creep from the site	Review at Design Qualification, design should only include elements required for XP515 manufacture	Project Manager	Medium
	Project change control process to prevent unnecessary change, limiting change to XP515 critical issues only	Review change control logs, rejected and approved request weekly during detailed design and prior to. Design Qualification	Project Manager	Medium
Aligned shutdown strategy	Project access requirements agreed and aligned with design requirements with no change to project scope objectives	Approval of design and project schedule by site production group	Project Manager	High
	Shutdown planning to include suitable contingencies and buffers	Review prior to construction start	Project Manager	Medium
Business case delivery	Design, procurement and construction completed on time	Construction completion dates as defined by project schedule	Project Manager	High
		Project handover completed as defined by project schedule	Project Manager	High
	Production capacity and stock build up for launch	Capacity 75% of design for first 3 months and 95% of design thereafter	Production Director	High

Following completion of these two tools it quickly became evident to Fred, the Project Manager, that there were still significant gaps in understanding who the project stakeholders were and their responsibilities to the project. These included:

- The known issues around firm identification of sponsor and customer.
- Issues around technology transfer from clinical trial and product development information into basic design requirements for the Engineering Contractor.
- Issues around operator training and commitment to quality assurance support staff training.

Fred called on the expertise of the Senior Process Engineer from the corporate engineering group to support resolution of these issues. The Senior Engineer was then seconded to the project for the remainder of the front-end design phase. Following his arrival at site the Project Team completed a Stakeholder Plan (Table 7-7) and Skills Matrix (Table 7-8) in order to identify responsibilities and gaps in skills and knowledge. Once gaps had been identified, additional personnel were brought into the project either from corporate engineering or the Engineering Contractor.

Table 7-7 Excerpt from Project Stakeholder Plan

***Project Management Toolkit* — Stakeholder Management Plan**						
Project: XP515 Upgrade				**Date:** Month 4		
Individual stakeholder analysis						
Stakeholder name/role	**Type**	**Current project knowledge**	**Level of influence**	**Current level of engagement**	**Target level of engagement**	**Management of stakeholder**
Gemma Farnsworth (Site Director)	S	Low	High	Low — only now identified as the real project sponsor as she is the person who can ensure that training and preparation issues for manufacture of XP515 are achieved	High — she must ensure the buy-in of all of her management team in order to ensure successful implementation, start up and ongoing production	Project Manager
Bob Smith (Production Director)	C	High	Medium	Medium — he is interested in the project both in terms of what he thinks he can get out of it and how far he can improve his facility. Production KPIs are still limiting his buy in to project shutdown requirements	High — need him to be fully engaged in the scope of the project as he will deliver the XP515 product following construction completion	Project Manager

(continued)

Table 7-7 (Continued)

Project Management Toolkit — Stakeholder Management Plan

Project: XP515 Upgrade	**Date:** Month 4

Individual stakeholder analysis

Stakeholder name/role	Type	Current project knowledge	Level of influence	Current level of engagement	Target level of engagement	Management of stakeholder
Gerry Badger (Quality Manager)	A	Low	Medium	Low — currently has little or no knowledge of the project and does not understand the likely impact on his department	High — in-process testing and final product testing will be critical to the manufacturing process and significant changes will be required to both testing capacity and complexity requirements. He will also be responsible for liaison with regulators	Project Manager
Barbara Booth (Engineering Manager)	A	High	Medium	Low — views the project as the way to get some new equipment into the plant and to generally make improvements	High — has responsibility for utilities and capacity issues which need early resolution in the project. Will ultimately have responsibility for equipment maintenance and will therefore will be part of handover approval	Project Engineer
Graham Roberts (Technical Services Manager)	A	Medium	Medium	Medium — is aware of the project and knows that his personnel will support the commissioning and qualification effort and then provide long-term plant support	High — his department will provide the bulk of the labour for qualification activities, plus the data to support effective technology transfer	Project Manager
Jim Kelly (Automation Manager)	A	High	High	High — he is aware of the project and the scope requirements. The project is included in his forward planning	High — his department has responsibility for a major element of project project scope	Project Manager

Summary stakeholder analysis

There are several issues that will have a likely impact on the overall success of the project. Although these are still to be resolved the stakeholders responsible for them are now identified which will allow progress and resolution to be achieved

Table 7-8 is an excerpt from the full analysis and has been included to highlight responsibility issues already identified within the project, for example shutdown and site preparation for XP515 manufacture.

Table 7-8 Excerpt from project skills matrix

	Bulk manufacturing	Bulk API manufacturing	Project management	cGMP	Automated systems development	Automated systems qualification	Clean room operations experience
Fred Jones (Project Manager)	✓	✓	✓	✓			
Bob Smith (Production Director)	✓						
Jim Kelly (Automation Manager)	✓		✓		✓		
Kevin Mackay (Automation Engineer)	✓		✓		✓		
Tom Banks (Quality Assurance)	✓	✓		✓		✓	
Gary Mather (Technical Services)	✓	✓		✓			✓

An example where a gap in the required skills has been identified is in the application of automated systems qualification. This is a heavily documented process with preparation required before the start of control system specification and configuration. The Quality Assurance Manager has experience of review and approval of systems at other sites, but resources with appropriate experience will have to be brought into the automation group for this project.

End of Stage Two situation

At the end of this phase (approximately 6 weeks duration) a capital request for $12.2 million was approved based on start of production at end of month 18. Following approval there are no further formal stage gates which the project must progress through. This approval is 'approval to deliver the facility as per the capital request'.

The external engineering company is developing a detailed programme based on the preliminary cost plan and schedule and intends to use earned value progress measurement where possible (Table 7-9).

At the end of this stage Fred conducted a review of the finalized project delivery plan through use of the "How?" Checklist (Table 7-10). This gave him reassurance that a robust plan was in place and that project delivery could commence. It also highlighted areas where further planning should be completed.

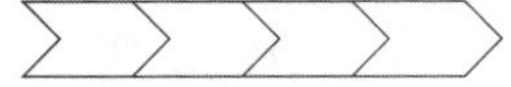

Table 7-9 Control Specification Table

***Project Management Toolkit* — Control Specification Table**		
Project: XP515 Upgrade	**Date:** Mid-Month 5	
Cost control		
Cost objective/CSF	**Control methodology**	**Responsibility**
Project budget must not exceed the revised cost of $12 million resulting from Engineering Contractor's conceptual design estimate	Engineering Contractor will provide cost/progress reports on an earned value basis throughout design and construction phases	Engineering Contractor/ Cost Controller
Schedule control		
Schedule objective/CSF	**Control methodology**	**Accountability**
Facility must be complete, handed over and operational by month 18	Engineering Contractor will maintain a Gannt-type schedule identifying schedule risks and potential overruns on a weekly basis	Engineering Contractor Scheduler
Scope control		
Scope objective/CSF	**Control methodology**	**Accountability**
Facility is approved by external regulators for commencement of manufacture with no non-conformances or additional requirements. Project scope must be limited to XP515 requirements	Control process design risks, all issues should be resolved prior to Design Qualification Manage change control	Quality Manager

Table 7-10 The 'How?' Checklist

<table>
<tr><th colspan="2">Project Management Toolkit — The 'How?' Checklist</th></tr>
<tr><td>Project: XP515 Upgrade</td><td>Project Manager: Fred Jones</td></tr>
<tr><td>Date: Mid-Month 5</td><td>Page: 1 of 4</td></tr>
<tr><td colspan="2">Stage One check</td></tr>
<tr><td colspan="2">Any changes since Stage One completion? (Development of the business case and project kick-off may be some time apart)
The project kicked off immediately following selection of the facility. No changes in the product strategy have been highlighted. The project has progressed through conceptual design & front end design and gained capital approval</td></tr>
<tr><td colspan="2">Sponsorship</td></tr>
<tr><td colspan="2">Who is the sponsor? (The person who is accountable for the delivery of the business benefits)
Gemma Farnsworth, Site Director.
Has the sponsor developed a communication plan?
Sponsor will meet with other key parties during normal weekly management meetings to receive feedback and to review project status</td></tr>
<tr><td colspan="2">Benefits management</td></tr>
<tr><td colspan="2">Has a benefits realization plan been developed?
A critical path has been identified (Figure 7-4) along with a Table of Critical Success Factors (Table 7-6) which outlines the project deliverables to allow realization of the benefits from the project
How will benefits be tracked? (Have they been adequately defined?)
See Table of Critical Success Factors</td></tr>
<tr><td colspan="2">Business change management</td></tr>
<tr><td colspan="2">How will the business change issues be managed during the implementation of this project? (Any specific resources or organizational issues?)
The Production Manager (Bob Smith) needs to understand the magnitude of the business change issues and develop a change plan
➤ Areas of the revamped facility will be operating to significantly different procedures than other areas
➤ There will be different access arrangements due to the change in facility segregation — for both operation and maintenance
➤ Some of the operator interface with the control system may be different
Quality, automation and technical services groups will be supporting the future operation of the revamped plant which will require a change in attitudes and practices to maintain the qualified state of the plant
Have all project stakeholders been identified? (Review the stakeholder map from Stage One)
Although not identified at the start of this project phase, stakeholders have now been identified and gaps around their roles and responsibilities are being dealt with by the Project Manager
What is the strategy for handover of this project to the business? (Link this to the project objectives)
The project will be handed following completion of operational qualification — this is normal practice</td></tr>
</table>

(continued)

Table 7-10 (Continued)

<table>
<tr><th colspan="2">Project Management Toolkit — The 'How?' Checklist</th></tr>
<tr><td>Project: XP515 Upgrade</td><td>Project Manager: Fred Jones</td></tr>
<tr><td>Date: Mid-Month 5</td><td>Page: 2 of 4</td></tr>
<tr><th colspan="2">Scope definition</th></tr>
<tr><td colspan="2">Has the scope changed since Stage One completion? (Has further conceptual design been completed which may have altered the scope?)
Some further conceptual design has been completed and an improved cost estimate developed.
Have the project objectives been defined and prioritized? (What is the project delivering?)
There is no indication that they have changed since conceptual design:
➤ Achieve the launch timetable
➤ Ensure that the facility is qualified so that the process can be validated and the regulatory approval obtained
➤ Minimize negative impact on plant utilization KPI. This may require negotiation in order to achieve other project objectives. See Table of Critical Success Factors</td></tr>
<tr><th colspan="2">Project type</th></tr>
<tr><td colspan="2">What type of project is to be delivered? (For example engineering or business change)
Capital engineering project
What project stages/stage gates will be used? (Key milestones, for example funding approval, etc., which might be go/no go points for the project)
The project was given pre-sanction funds for development of the design and then this was used to develop the final capital funding authorization. Following capital approval the company has no further formal stage gates</td></tr>
<tr><th colspan="2">Funding strategy and finance management</th></tr>
<tr><td colspan="2">Has a funding strategy been defined? (How will the project be funded and when do funds needs to be requested?)
Yes — this is a standard process which is being followed
How will finance be managed?
The Cost Controller and Project Manager within the Engineering Contractor team will report to the site Project Manager on a weekly basis</td></tr>
<tr><th colspan="2">Risk and issue management</th></tr>
<tr><td colspan="2">Have the critical success factors changed since Stage One completion? (As linked to the prioritized project objectives and the critical path through the project risks)
No
Have all project risks been defined and analysed? (What will stop the achievement of success?)
The current issues have been outlined and some mitigating actions are in progress, but there is a lack of prioritization of these actions; there is no analysis of which issues would have the greatest impact on the achievement of project critical success factors
What mitigation plans are being put into place?
See above
What contingency plans are being reviewed?
None at present</td></tr>
</table>

(continued)

Table 7-10 (Continued)

Project Management Toolkit — The 'How?' Checklist

Project: XP515 Upgrade	**Project Manager:** Fred Jones
Date: Mid-Month 5	**Page:** 3 of 4

Project organization

Who is the Project Manager?
Fred Jones (allocated by the corporate engineering group)
Has a project organization for all resources been defined? (Include the Project Team and all key stakeholders)
Yes (Figure 7-3), but there are still issues remaining within the organizational structure including:

- Further work with sponsors to ensure their understanding of the project and their requirement to provide resources for specific activities
- Additional automation department resource required to bring automation qualification experience to the group
- Links to be formed to obtain design information from clinical trials team and NDA data. Senior Corporate Process Engineer has already been seconded to the Project Team until issues are resolved

Contract and supplier management

Has a strategy for use of external suppliers been defined? (The reasons why we would need to use an external supplier for any part of the scope)
Yes — the project will use an external Engineering Contractor. This company will produce a procurement plan for the engagement of all other external suppliers (equipment vendors and installation subcontractors)
Is there a process for using an external supplier? (selection criteria, contractual arrangements, performance management, etc.)
Engineering Contractor was selected by corporate engineering using their procedures

Project controls strategy

Is the control strategy defined?
Only at a high level (Table 7-9) — the majority of the detailed control of the project will be the responsibility of the external Engineering Contractor:

- Cost control strategy — external cost engineering linked with Internal Project Manager
- Schedule strategy — external planner linked with Internal Project Manager
- Change control — not stated
- Action/progress management — earned value
- Reporting — monthly project report

Project review strategy

Is the review strategy defined? (How will performance be managed and monitored — both formal and informal reviews and those within and independent to the team?)
Yes — The Project Manager will meet with the Production Manager every 2 weeks, Production Manager will communicate with other key stakeholders at existing weekly management meetings

(continued)

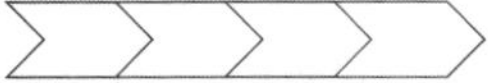

Table 7-10 (Continued)

Project Management Toolkit — The 'How?' Checklist	
Project: XP515 Upgrade	**Project Manager:** Fred Jones
Date: Mid-Month 5	**Page:** 4 of 4
Stage Two decision	
Should the project be progressed further? (Is the project delivery strategy robust enough for project delivery to commence?) Yes — although there are some gaps which need to be addressed immediately: ➤ Completion of stakeholder management ➤ Risk management — need to prioritize risks so that mitigation actions can be appropriately prioritized ➤ Contract strategy — an outline strategy should've been given to the external engineering company. This would have provided a basic translation of the project CSFs into contract strategy ➤ Control strategy — complete reliance on the engineering company is inappropriate as they are only managing to mechanical completion — the PM is responsible for delivering a qualified facility ready for operation	

Lessons learnt

A brief after-action review highlighted the following key lessons learnt:

- All stakeholders and specifically the project sponsor should have been identified earlier in the project. However, this exercise has now been completed with little detrimental effect on the project allowing resolution and management of the issues.
- Additional requirements for roles and responsibilities within the project have now been identified. These included an including Automation Engineer with qualification experience and secondment of the Corporate Senior Process Engineer to ensure the availability of design information for technology transfer. The project has been lucky that a Corporate Process Engineering resource could be made available. This could have had serious implications for the project as the Engineering Contractor may have been progressing inappropriate designs which would have required significant rework at a later stage in the project, incurring time and cost penalties.
- Plant utilization targets may have to be renegotiated in order to achieve project benefits. The Project Manager can now work with the Project Sponsor, the Production Director and the Site Director to address this issue and determine priorities for the project and for the site. This will help to define and agree requirements for access to the plant and for the actual shutdown period.

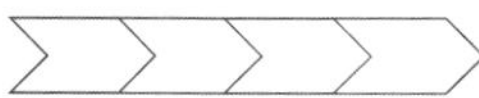

Stage Three — Project Delivery

A review of the project delivery phase is provided along with examples of each of the "In Control" Tools that were used during the phase. Each tool was actually used several times during the delivery phase to assess progress and to provide forecast project costs and end dates. They were also used as reporting tools to the project stakeholders.

Design

The detailed design commenced immediately following capital funds approval in the middle of month 5. During the early part of month 6 a review of the sales forecasts indicated that a check of the design was needed. The design team proposed increased automation as the method to increase overall capacity (to try to reduce the cycle time) and this was approved.

It is currently the end of month 7 and apart from detailed design additional planning work is in progress including the development of the Validation Master Plan with associated validation matrix.

The facility has been operating in its present state for 7 years, having previously been updated and re-engineered to manufacture the current product requirement. Elements of the original plant are approximately 15 years old. The design engineers haven't been in the plant and are relying on the documentation issued to them.

Bob Smith (Production Director) has lost some interest in the project. This is partly due to the requirements of his normal manufacturing role and stock build ups to allow for plant shutdowns, and partly as he has realized he has very little influence over the project scope and additions he would like to make. At present Fred, the Project Manager needs him to focus on:

- The latest date for the shutdown to commence.
- The earliest date for re-start.
- The training needs for his staff.
- Likely staffing levels, reviewing whether any changes seem likely.
- The new maintenance needs of the facility.

The basic facility layout information will be discussed with Bob as the design progresses, as a method to engage him. Current concerns are:

- Bob's manufacturing operation will be impacted pre-shutdown due to the level of resource we would like in the Project Team. It will then be impacted during the shutdown as all areas except utility generation will be handed over to the project.
- Post-shutdown Bob will be running a more flexible facility with the potential to be available for future new APIs. This project will effectively secure the medium-term future of this manufacturing facility, which was in doubt before the project was announced.
- Bob remains concerned by the reduction in his plant utilization, but this is understandable considering the climate he has been working (high utilization = high potential to keep the facility within the company). The Site Director is working to readjust Bob's annual targets. However the agreement on shutdown requirements is not fully resolved with the project requiring 6 months and Bob only agreeing to 4 months.

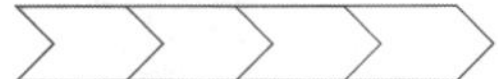

The detailed design is approximately 50% complete (earned value against a 60% actual and 50% planned), but a GMP review raised a number of issues with the facility layout and basic process design:

The project risk assessment was reviewed (Table 7-11) and the likely achievement of the Critical Risk Factors was determined (Figure 7-5).

Table 7-11 Risk Table

Project Management Toolkit — Risk Table						
Project: XP515 upgrade			**Date:** End Month 7			
Risk description			**Risk assessment**		**Action planning**	
Risk No.	**Risk description**	**Risk consequence**	**Occur?**	**Impact?**	**Mitigation plan**	**Contingency plan**
CSF1 — Strategic alignment						
1.	Several raw materials from other processes have been found to be incompatible with XP515 manufacture	Batch contamination and unwanted side reactions during process step	Low	High	A new solvent based cleaning process in place of water/detergent based will be introduced into the facility following use of these materials for all future plant 2B production operations	Not applicable
CSF2 — Robust technology transfer						
2.	Equivalency of API isolation mechanism not proven	Regulatory approval problems and delays	High	High	A plan for the development of scientific proof of equivalence must be developed by Engineering Contractor and Product Development group	Alternative and more expensive engineering solutions and equipment to be reviewed and costed. Initial discussions with suppliers to take place
CSF3 — Facility qualification and cGMP compliance						
3.	Availability of vendor documentation at project handover	Delay of handover, impact on development of maintenance programme and engineering records	Medium	Low	Engineering Contractor and document controllers will prepare a matrix of all documentation that has been requested. Report of requested versus received will become part of the normal project report. Vendors will be asked to supply documents under separate cover direct to the Engineering Contractors Offices	N/A
4.	Delays during qualification process impacting overall schedule	Delay of plant handover for operation	Medium	Medium	Once detailed design works are complete designers will meet with site technical operations to agree on commissioning and qualification requirements and schedule	N/A

(continued)

Table 7-11 Risk Table (Continued)

Project Management Toolkit — Risk Table

Project: XP515 upgrade | **Date:** End Month 7

Risk description			Risk assessment		Action planning	
Risk No.	**Risk description**	**Risk consequence**	**Occur?**	**Impact?**	**Mitigation plan**	**Contingency plan**
CSF4 — Appropriate scope definition						
5.	Scope creep due to additional requirements from site coming from process and technical reviews	Minimal as a rigorous change control process is in place	Low	Low	N/A	N/A
CSF5 — Aligned shutdown strategy						
6.	Shutdown requirements have not been fully agreed with Production Director. Project sponsor is now managing this issue. There are issues around materials that cannot be manufactured in 2B if project shutdown requirements are fully met	All required works will not be completed within the given time	Medium	High	Project sponsor to resolve this issue. If this issue cannot be resolved directly then several other options may be open: – Source the required materials from an external source – Consider the impact of not supplying the materials	This issue must be resolved by the end of month 8 and before the start of procurement. Project Manager is to meet with Sponsor and Production Director to ensure this issue is resolved
CSF6 — Business case delivery						
7.	Detailed design is indicating a potential for increase in capital cost. Additional clinical trial product information has required improvement to some equipment and area finishes	Project cost and plant bottom line will be impacted	Low	Medium	Project Manager will control project spend within existing contingencies	N/A
8.	There are schedule issues around two major equipment items both of which will arrive extremely late in the overall installation programme	Any delay in these items will impact product launch	High	High	Engineering Contractor is aiming to release final specifications for these items before all others. Items are single source and initial discussions with vendors have already taken place. Regular vendor tracking will take place and all document approvals will be walked round by hand for approval within 2 working days	Cost bonuses or liquidated damages may have to be considered for these items

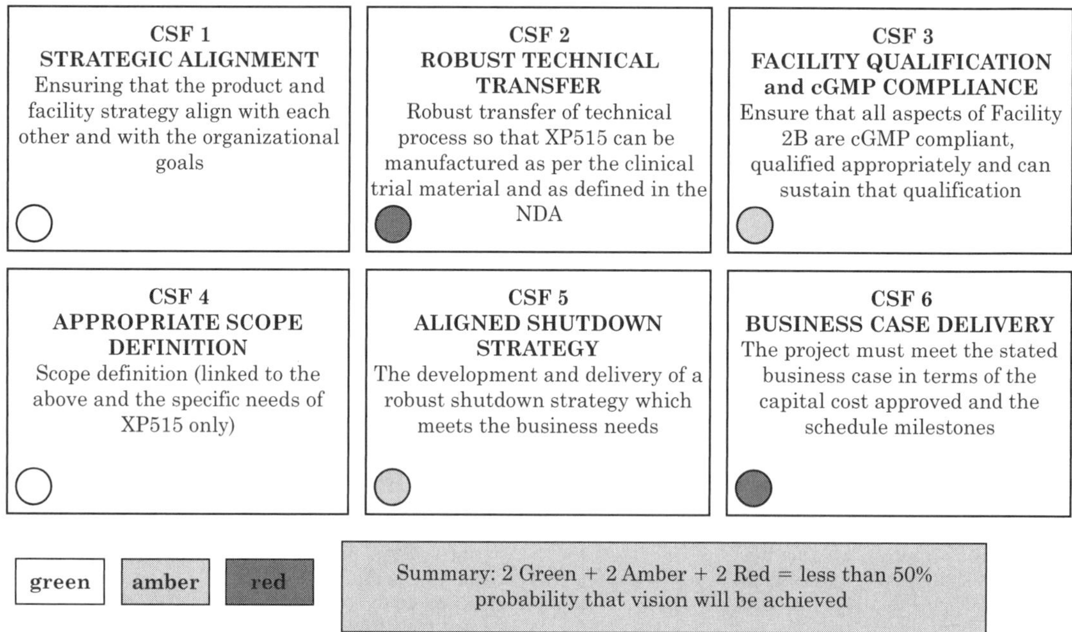

Figure 7-5 Path of success — progress

The Milestone Progress Report was produced monthly as a high-level reporting tool for the site alongside the completion of the normally required monthly project reports (Table 7-12). At the end of the month 7 the progress report was able to demonstrate that key milestones were being met and that the design and procurement activities were interfacing appropriately.

Table 7-12 Project Milestone Progress Report

***Project Management Toolkit* — Milestone Progress Report**				
Project: XP515 upgrade			**Date prepared:** End Month 7	
Report prepared by: Fred Jones			**Schedule reference:** X515-M7	
Milestone progress				
Ref	**Activity**	**Planned date**	**Actual or forecast date**	**Comment**
1.5	Overall Project risk assessment	End Month 6	End Month 6 + 2 weeks	Meetings were delayed to allow for half-term school holidays and missing personnel. No negative impact on project
2.3.1	Complete specifications for early order of major equipment items	Month 7	Month 7	Engineering Design Contractor has achieved schedule requirement. Procurement group to complete item purchase by month 8
2.3	Completion of detailed design	Month 9	Month 9	Procurement specifications will be released in batches to procurement department. The impact of this is preferable to procurement and the schedule for release has been discussed internally. No impact to the project is expected
3.1	Commence equipment procurement	Month 9	Month 9	See comment for 2.3 in the above row
3.2	Commence installation contract procurement	Month 10	Month 10	N/A
4	Construction	Months 12–18	Months 12–18	
	Completion date	Month 18	Month 18	Month 18

Ref: The reference number is taken from the main project schedule/work breakdown structure.

Procurement

Procurement started at the end of month 9 with the exception of the two main plant items which required early order. Corporate Engineering were insistent that corporate procurement guidelines were followed throughout the project, however this started to impact project schedule. Eventually it was agreed that deviation from the standard process of tendering to three suppliers for each separate item would be relaxed on a case-by-case basis.

The external engineering company was contracted to plan and deliver the procurement of all items in line with the needs of the project. Their procurement strategy was as follows:

- Use the equipment listing to develop the list of procurement packages.
- Procure two main plant items early in the schedule from single vendors. The equipment is proven for the process and pilot-scale versions were used in the manufacture of clinical trial materials. They also have extremely long delivery times. This strategy was agreed with Corporate Engineering.
- Tender to three companies for all items to get competitive prices.
- Aim to have a fixed priced contract for all items to manage risks related to the project budget, in line with Corporate Guidelines.
- Don't use any one company for more than one procurement package.

The procurement progressed according to plan but the requirement to always have a fixed-price contract caused some issues:

- The tender documents and associated technical specifications had to be absolutely complete and require little or no supplier input. This impacted schedule as in some cases tender packages were issued late.
- The costs bid by suppliers were higher than anticipated due to their contract style which included built in cost contingency.

During month 10 Fred, the Project Manager finally got agreement with Bob, the Production Director that the facility would be shutdown in month 14 for 4 months, with a modified shutdown programme to allow project works to progress with the plant remaining operational. This has required a change in the project scope as additional temporary segregation for the pre-shutdown installation works will have to be completed at increased project expense. A project change request for additional funds has been submitted, supported by the business case for not shutting the plant down for more time. The change has followed the project change control process. The project sponsor has approved the need for the change, identified additional funds and agreed to progress it through the project approval process. The change does not have a schedule impact on the project.

Fred continued to check project progress on a regular basis. At the end of Month 11 he concluded that there were some issues within the procurement phase and that these were mainly caused by the requirement to adhere to corporate guidelines which didn't align with the project delivery strategy (Table 7-13):

(a) the project schedule needed to buy long lead items as quickly as possible but this was delayed by 2 weeks due to additional scrutiny at the corporate level even though the procurement strategy had been approved.

(b) in order to meet the critical path some installation contracts would be better managed as target reimbursable, thus allowing the release of design in phases, and commencement of the installation works earlier.

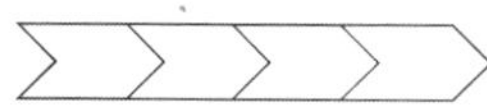

Table 7-13 'In Control?' Checklist

Project Management Toolkit — The 'In Control?' Checklist

Project: XP515 Upgrade	**Project Manager:** Fred Jones
Date: End Month 11	**Page:** 1 of 3

Stage Two check

Any changes since Stage Two completion?
Yes — The length of the shutdown for the project has now been agreed for 4 months starting in month 14. This has required a change in construction strategy to allow concurrent operation of the plant. Project change control has been used and a project change request has been submitted for approval. Gemma Farnsworth, the project sponsor, will progress the change through off-site management approval

Business change management

What is the current status of stakeholder management? (Review the original stakeholder map and discuss)
Weekly communication to the project stakeholders takes place at the site management meeting. As the Site Director is the sponsor for the project she ensures that time is made within the meeting to discuss the project's progress, current requirements and forward plan
How will the business be expected to operate as a result of the completion of this project?
The project will impact the business in the following ways:

- Existing plant operators will operate the new plant but will require new training and skills especially in the areas of containment, new changing regime and product sampling
- Quality control will see increased testing requirements during XP515 manufacture. A new piece of equipment is required for one of the in-process tests, the required equipment has been sourced from another site. Lab technicians require training which will be provided by the equipment manufacturer
- Technical services will provide process support to the XP515 process. They have already identified the personnel within the department who will take responsibility for XP515. They have been involved in process reviews and GMP reviews so have developed links with the team who manufactured the clinical trial materials

Is the business ready for this project?
Plans are in place which will prepare the business for completion of the project. Progress is reviewed at the weekly site management meeting as part of the project discussions
What is the strategy for handover of this project to the business?
The project will be handed over following completion of commissioning, qualification and close out of agreed issues

Scope definition

Has the scope changed since Stage Two completion?
Yes. — see notes above relating to shutdown requirements
What is the project progress against the Stage Two defined and prioritized project objectives?
The two main early order equipment items were not procured on schedule as single sourcing is not compliant with site purchasing department procedures. The purchasing department blocked the approval of the orders until they had provided their own level of scrutiny. This delayed the order of these two items by 2 weeks which impacted the construction schedule by 2 weeks (critical path issue). Additional cost and schedule issues were caused by the corporate requirement to have fixed price contracts

Project roadmap

What project stages/stage gates have been completed? (For example project approved, design complete)
Project is fully financially approved; detailed design is complete

(continued)

Table 7-13 (Continued)

Project Management Toolkit — The 'In Control?' Checklist

Project: XP515 Upgrade	**Project Manager:** Fred Jones
Date: End Month 11	**Page:** 2 of 3

Risk and issue management

Have all project risks been reviewed regularly during project delivery?
Project risk assessment was reviewed at the end of month 7 and thereafter at monthly intervals
What is the status of mitigation plans?
Isolation step equivalence — data collection is complete and final report is nearing completion
Procurement process requested delivery of vendor and contractor documentation under separate cover to provide improved control over documentation
Commissioning and qualification plan is 75% complete
Shutdown requirements have been agreed by all major stakeholders. If additional funding request is not approved then plant will be shutdown for longer with the consequent loss of product understood & agreed to by the project sponsor and the Production Director
What is the status of contingency plans?
Contingency plans have not been required at this point
What is the overall likelihood of achieving the project critical success factors?
<50%, as there are issues around schedule for the late delivery of the two main equipment items. Construction plan is to be reviewed in order to find a way of keeping to schedule and allowing a later installation of the equipment

Project organization

Are project activities being completed by the appropriate members of the organization?
Project Team is functioning effectively. Document approvals outside of the Project Team have taken on average 1 week longer than anticipated and this issue has been identified at weekly stakeholder meetings

Contract and supplier management

What external suppliers are being used?
All construction activities other than control system installation and software development will be provided by external installation Contractors: mechanical, electrical and instrumentation, civil (major works and external), clean room specialist, fire protection
What is the external supplier status and performance?
Mechanical equipment supply is progressing as planned
Temperature transmitters will be delivered late, this can be managed within the schedule
Civil contractor will commence external works in 2 weeks (not within the shutdown area)

Project controls strategy

Are project costs under control? (Review cost plan — actual vs budget, etc.)
Yes — project cost is under control
What is the likelihood that the project budget will be maintained (forecast to completion?)
Project is expected to remain within allowable contingencies
Is the project schedule under control? (Review schedule and milestone progress, etc.)
Schedule is under control (with some internal delays not impacting the critical path and one external delay which does). Reporting process from Engineering Contractor to Project Manager and stakeholders is working
What is the likelihood that the project schedule will be achieved (forecast to completion?)
Forecast completion is currently two weeks late due to major equipment delivery delays (late order placement)
Are there any changes to scope (quantity, quality and functionality)? Are the costs and schedule under control?
Yes — change control has worked throughout the design & procurement process

(continued)

Table 7-13 (Continued)

<table>
<tr><th colspan="2">*Project Management Toolkit* — The 'In Control?' Checklist</th></tr>
<tr><td>**Project:** XP515 Upgrade</td><td>**Project Manager:** Fred Jones</td></tr>
<tr><td>**Date:** End Month 11</td><td>**Page:** 3 of 3</td></tr>
<tr><td colspan="2">**Project review strategy**</td></tr>
<tr><td colspan="2">**Are regular Stage Three reviews being conducted? (Is performance being managed and monitored?)**
Stage Three reviews are being completed monthly
Is project performance adequate for project success?
Project Team is performing well, the procurement problem was unforeseen and must be included in any lessons learnt process at the end of the project
Is there regular reporting? (Is the Project Team adequately managing communication of progress and performance to all stakeholders?)
Yes — see previous comments on stakeholder management</td></tr>
<tr><td colspan="2">**Stage Three decision**</td></tr>
<tr><td colspan="2">**Is the project under control? (Is the project control strategy robust enough for project delivery to continue?)**
Yes — the project should progress
What is the certainty that the project will be successful?
Currently less than 50%. Schedule issues must be addressed and resolved</td></tr>
</table>

Construction

The majority of the facility was handed over late to the Project Team, halfway through month 14 (a product campaign had to be completed that was only scheduled 2 months before). The remainder of the facility is manufacturing an excipient AB2A that has been ordered at short notice by another manufacturing site.

The status at month 17 was as follows:

- A Safety audit conducted the previous week highlighted:
  - Different contracting companies working in the same area — some above the other.
  - Inadequate levels of scaffolding for the level of resource working in some areas.
  - Poor housekeeping due to lack of responsibility by contractors to clear up after themselves.
  - Insufficient control between construction and operational areas.
- A GMP audit conducted yesterday highlighted:
  - Contractors accessing production areas without due regard to the ongoing production — often they used shortcuts to access another area where it was OK to work.
  - Poor quality of work specifically in the mechanical installation.
  - Lack of current documentation for works being completed.
  - Concerns that the poor housekeeping would impact the operational areas of the facility.

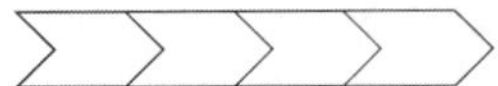

- The Project Manager walked around the site with the part-time Construction Manager and concluded that:
  - All areas of the services construction that are not dependent on the production activities are complete.
  - Installation of production equipment is 90% complete due to main equipment items which have not yet been delivered.
  - Mechanical installation commenced late (50% complete). The contractor has therefore doubled site resources and also used a part of the service area for fabrication in order to speed things up.
  - Electrical and instrumentation installation teams arrived on site on time but have been working inefficiently due to waiting for mechanical installation to complete (10% progress).
  - Not all on-site personnel appear to be working effectively as they are waiting for materials, access or permits to work, etc.
  - Control system integration and software development is well advanced. Testing and qualification are progressing well. There are no apparent issues.

The issues were:

- It quickly became apparent that the scheduling of site construction and handover cannot be achieved as existing production is now so critical that a complete shutdown of the facility is not possible. The completion of construction and handover has therefore had to be phased to meet this requirement.
- The phasing of works to allow for ongoing excipient AB2A production required very minor increases to project scope. The project management team have assumed that the overall scope of the project has not changed sufficiently to warrant any reconsideration of budget or schedule.
- The Engineering Contractor has focused his efforts to meet the phasing requirement, and as access to key production areas of the plant was limited he has concentrated on the completion of the services construction.
- The production construction was delayed by this as access to specific production areas has been limited due to ongoing production.

Commissioning and qualification

The final completion date for construction was 6 weeks behind schedule. This was due partly to the late delivery of main equipment items caused during procurement and partly due to the ongoing manufacturing that was carried on in plant until half way through month 16.

A re-planning exercise during the construction phase was unable to resolve the issues and the equipment manufacturers could not speed up their delivery programme due to their own production constraints.

A phased construction handover was therefore planned and implemented at the end of month 18 allowing the start of commissioning and qualification of fully complete systems. While this did not completely solve the schedule issues and bring the schedule back in line it did allow the installation and operational qualification to finish 2 weeks earlier than planned, limiting the overall schedule overrun to 4 weeks.

The following issues were noted:

- Care had to be taken to ensure that only complete systems were handed over. Commencing qualification of incomplete systems can generate a lot of non-conformances which require individual close out and approval.
- Marked up engineering drawings showing the extent of what was being handed over for each system were completed by the Construction Supervisor and issued to the Technical Services Department to communicate boundary limits.

The plant was complete and ready for start up by the end of month 19 (Table 7-14). Minor remedial works were required by installation contractors, none of which were classified as quality impacting. Regulatory inspection was completed early in month 20. The inspection was rescheduled following project reviews during month 17 when the re-planning exercise forecasted that the project completion would be delayed.

Table 7-14 Project Scorecard

***Project Management Toolkit* — Project Scorecard**	
Project: XP515 Upgrade	**Date:** Month 19
Project status summary	
Project status	Construction, commissioning and qualification are complete. The plant is ready to commence the manufacture of the process qualification batches. This is approximately 1 month behind schedule
Project control status	Risks to the project have focused on the schedule delay and its impact on the business case objectives
Key activities this period	**Achievements**
Construction, commissioning and qualification phase have been completed and plant handed over for performance qualification & regulatory inspection	6-week delay has been reduced to 4 weeks by phased handover process. The management of this has been onerous for the Project Team and required some over time working but has proved worthwhile
Key activities next period	**Critical issues**
Complete performance qualification Commence and complete operator training Complete lab technician training for new testing requirement	Completion of performance qualification Inspection and approval by external regulators

(continued)

Table 7-14 (Continued)

Project Management Toolkit — Project Scorecard				
Project: XP515 Upgrade			**Date:** Month 19	
Critical success factor tracking				
CSF level 1		**Progress**		**Risk rating**
CSF 1 strategic alignment	○ Green	Facility design is complete	○ Green	N/A
CSF 2 robust technical transfer	○ Green	Facility design is complete	○ Green	N/A
CSF 3 facility qualification and GMP compliance	○ Amber	Qualification and commissioning has commenced 6 weeks late in the overall schedule but in line with a phased handover to limit the impact of construction delays	○ Amber	N/A
CSF 4 appropriate scope definition	○ Green	Facility design is complete	○ Green	N/A
CSF 5 aligned shutdown strategy	○ Green	Construction is complete	○ Green	N/A
CSF 6 business case delivery	○ Amber	Project completion is 4 weeks later than planned Performance qualification must be completed on time The regulatory inspection must be completed on time with no major comments or change requirements (a risk review should be scheduled as soon as possible)	● Red	Risk review to be completed ASAP. The risk of any further delay to the project must be reduced as far as possible
Overall project	○ Amber	Project has incurred a 4 week overall delay due to production requirements. This additional production requirement was met and customers received their product as ordered. The delay has been communicated to senior off-site management	● Red	The project has been delayed by 4 weeks. Production should start as soon as possible. Sales and marketing are reviewing options for the quantities of stock build up prior to launch in order to try minimize any impact on product launch date. The site is reviewing the possibility for additional shift working to increase production capacity once plant is fully proven and process qualification batches have been produced

End of Stage Three situation

The project has been delayed by 4 weeks due to unforeseen production requirements. Strategies are under consideration for the limitation of the impact of this delay. However, forecast production start up and manufacturing dates indicate that the product will be launched 5 months ahead of its competitor and the company should be able to achieve the originally forecasted market share.

The project has also seen a small cost overrun: Additional time incurred due to the ongoing plant operations has meant increased contractor construction and mobilization costs due to the need for more flexible, less efficient working practices.

Lessons learnt

A brief after-action review highlighted the following key lessons learnt:

- Despite the delay the project has progressed extremely well and the delivery phase can be considered a success for many reasons:
  - The delay was outside the control of the project and did allow the manufacture of a raw material product. This was required for the production of another API product at another company site, allowing rapid turn around for a large Japanese order.
  - The project communication and reporting process ensured that the issue was well understood and the decisions taken were based on sound information. This also meant that contingency plans, both onsite and offsite, were considered at an early stage.
- The project did manage to recover 2 weeks of the original 6 week delay by phasing the project handover process.
- However the single source procurement issues which arose over the main equipment items should not have happened. Earlier meetings between the Project Manager and the site purchasing department would have resolved this issue and removed this delay.

Stage Four — Benefits Realization

Following mechanical completion of the project the customer and sponsors took responsibility for the start up, operation and maintenance of the facility including the provision of all necessary technical services and quality support.

They were not familiar with sustainability planning but with support completed the following activities:

- Benefits tracking
  - See Table 7-15 for the Benefits Tracking Tool as at month 22.
  - Note that at this stage not all benefits had been realized, so tracking should continue.
  - The operations team have already decided that manufacturing cost and capacity will continue to be measured as key operational performance measures.
- Sustainability checking
  - See Table 7-16 which shows the first of these reviews.
  - The sustainability checks are the 'early warning signs' that the changes are not being sustained sufficiently to realize the full set of benefits.
  - At this stage it looks like there are some issues with the plant operations and a plan is to be put in place to investigate. If the plant does not operate as intended then it will not be able to achieve the cost and capacity needs as required by the marketplace. This may affect launch stocks and will certainly affect longer-term post-launch supply if not rectified.
- Overall benefits check
  - See Table 7-17 which shows the Benefits Realized Checklist.

Table 7-15 Benefits Tracking Tool

Project Management Toolkit — Benefits Tracking Tool

Project: XP515 Upgrade			**Date:** Month 22			
Benefit metric		**Baseline**	**Milestone 1**	**Milestone 2**	**Milestone 3**	**Target**
		Month 20	Month 21	Month 22	Month 23	Month 26
Metric 1: Design, procurement and construction schedule completion dates	Plan Actual	Month 18 Month 19	N/A N/A	N/A N/A	N/A N/A	N/A N/A
Metric 2: Facility approved by external regulators	Plan Actual	Month 19 Month 20	N/A N/A	N/A N/A	N/A N/A	N/A N/A
Metric 3: Production capacity (% of design)	Plan Actual	75% 33%	75% 78%	75% 82%	95%	95%
Metric 4: Manufacturing cost ($ per kg)	Plan Actual	$250 $350	$250 $295	$210 $230	$210	$210

Table 7-16 Sustainability Checklist

Project Management Toolkit — Sustainability Checklist

Project: XP515 Upgrade	**Date:** Month 22

Project vision

XP515 must reach the market before the competitor — secure market share
Maintain production capacity, quality standards and product costs — maintain market share

Sustainability review information

Previous sustainability review: This is the first review	**This sustainability review:** Month 22
Project representative: Fred Jones (Project Manager)	**Customer representative:** Bob Smith (Production Director)

Sustainability checks

Check No.	Check	Target (sustained change)	Last review	This review
1.	Plant is operating as intended by design. Number of process deviations per batch	0 major 1 minor	N/A	2 major 4 minor
2.	Plant is operating as intended by design. Clean area monitoring records clean area compliance during operation (audit issues)	0	N/A	2
3.	Between campaign cleaning for different products is completed as planned. All cleaning swabs pass technical services testing	All	N/A	N/A
4.	XP515 product meets testing requirements and does not cause process delay. Final batch release is no longer than 36 hours	0 delay 36 hours	N/A N/A	6 24 hours

Summary comments and next steps

Improvement plan is in place for plant operations. Cleaning will be measured at the end of this XP515 campaign and prior to the next

Is the change completely sustained?	~~Yes~~/No	**Date of next sustainability check**	Month 23

Table 7-17 Benefits Realized Checklist

Project Management Toolkit — The 'Benefits Realized?' Checklist

Project: XP515 upgrade	**Project Manager:** Fred Jones
Date: Month 22	**Page:** 1 of 1

Stage Three check

Any changes since Stage Three Completion? (Note only the changes since the final Stage Three 'health check')
No changes since last Stage Three health check

Business benefits

Has the business case changed since Stage One? (For example during planning and delivery, pre- or post-project approval)
The business case has not changed; however, a project change which impacted the project's ability to meet the planned business case did take place. The decision to not shut the plant down as planned caused an overall delay to the project of four weeks and a cost overrun. Sales and marketing have altered their strategy for pre-launch stock build up and shifted the product launch date by 2 weeks

Have all benefits been defined in terms of trackable metrics? (Why is the project being done?)
Clear benefit metrics have been specified

What is the customer feedback? (Feedback from all stakeholders in the business including the customer)
The project delivery is considered a success by stakeholders. Communication and reporting process has worked extremely well

Are the benefits being tracked?
Yes — benefits are currently being tracked (Table 7-15) although final targets have not yet been achieved

Business change

Is the business ready for this project? (If the project can only enable benefits delivery by changing the way people work — has this been delivered, for example training?)
Operator training was included in the project plan and was complete prior to plant start up

Quality assurance lab technicians were trained in the use of new equipment and procedures Sustainability checks highlighted that additional business change actions may be needed (Table 7-16)

Scope definition

Has the scope been delivered?
Yes — the project scope has been delivered and the plant is on track to meet production capacity requirements

Have the benefit enablers been delivered? (Are you sure that the project will enable the benefits to be delivered now the project is complete?)
Yes

Stage Four decision

Has the project been delivered? (Delivery of project critical success criteria)
Yes

Have the business benefits been delivered? (Why was the project done in the first place?)
No — not all business benefits have been fully achieved. The issues have been communicated and mitigation plans are in place to minimize the impact on the product launch

End of Stage Four situation

The project was completed and the new product successfully introduced and launched. Initial sales exceeded forecasts and the facility was on-line 2 months ahead of the date forecasted by the Spanish manufacturing site. The project was therefore successful.
The project did not achieve 'typical' project success criteria:

- The project was handed over to the operations team for performance qualification 4 weeks late.
- The final project cost was $12.5 million. This represents a 2.5% overspend, requiring no additional capital approval as it was within the 10% contingency allocated to the project.
- Production costs are reducing but still over the design value. Costs should decrease as full plant capacity is achieved.

Lessons learnt

A brief after-action review highlighted the following key lessons learnt:

- Project plans can be flexible if supported by robust project processes; communication, reporting; stakeholder management and risk assessment/management. These processes have been key to making this project a success rather than one which did not deliver on time. Because all stakeholders were aware of the risks and consequences to the project as it progressed they were able to enact their own contingency plans to achieve the overall goals of the project and the company.
- Project success can only be measured by considering the overall needs of the business thus supporting the previous statement in the book that 'no project is an island.'

Conclusion

Capital engineering projects have a very fixed and mature project roadmap. It links together the way funds are authorized and managed with the way that compliant facilities should be designed and built.

The very nature of the information flow around the roadmap, particularly during the design stage, makes these projects complex. New ways of working are needed if control of all aspects is to be maintained: cost, time, quality, quantity and functionality of scope.

- Design information is needed for procurement, but equally procurement information is needed to complete the design.
- The design should be complete before the installation occurs. The installation should be complete before testing occurs. Both situations lead to complex phasing of various parts of a facility project.

Figure 7-6 highlights how different organizations have approached these issues. Some manage the complex iteration of information flows whilst others challenge the logic links and break the facility into its constituent parts: a building and a process. In the latter situation the benefits seen have been:

- No functional behaviours between different engineering disciplines or different parts of the Project Team (client vs contractor, etc.).
- Compliance is 'built in'.
- Elimination of 'handovers'.
- Well-managed interfaces.
- Information 'flows'.
- More agile decision making.

The project management challenge for pharma facility projects is to maintain a focus on qualification and compliance *and* on project management rigour. Fortunately the latter supports the former.

Points to remember

- Capital expenditure, and associated capital funding approval, is usually a very sophisticated procedure which is specific to a particular organization.
- Pharmaceutical facility projects have additional constraints as they must 'sit' appropriately within a highly regulated environment. They have to comply with regulations which are there to ensure the safety of the drugs made in the facility.
- Pharma facility projects usually have clearly defined schedule drivers due to the competitive nature of the industry and the drivers from patients, e.g. getting new drugs to the market and to patients has the ability to save and/or enhance lives of those that need the drugs.
- Pharma facility projects have the same basic four stages as any generic project and as such the tools in this toolkit are completely applicable. Additionally they can be used to manage the compliance and regulatory issues which are a feature of this type of project.

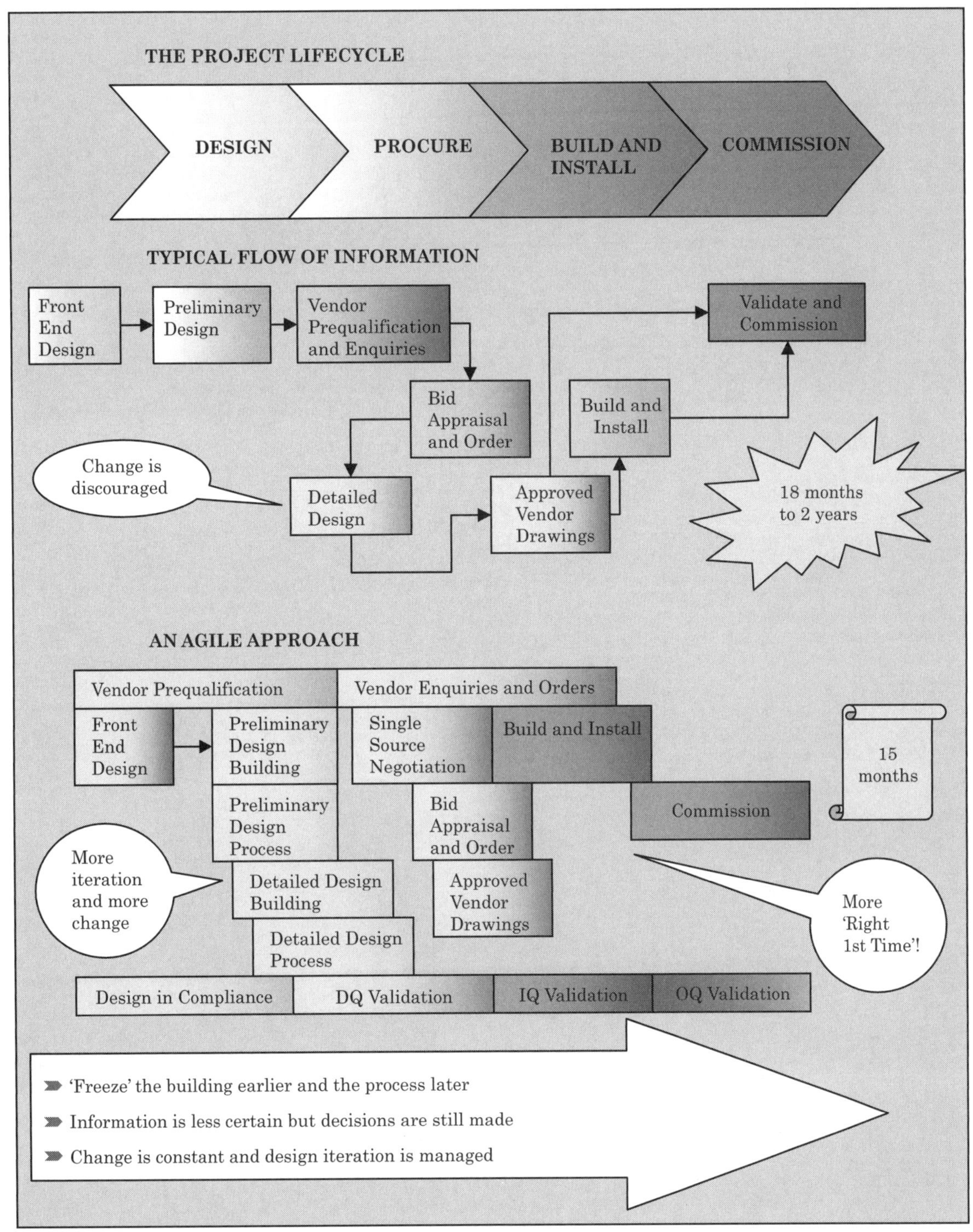

Figure 7-6 Project delivery information flows

8 Case Study Two: the business change project

This case study is taken from a real business change project recently and successfully implemented. It has been used for a number of reasons.

1. *Project type*

This is a business change project. Such projects are delivered in direct response to an organizational 'issue' or to better align a business area with an organization's strategic goals. These projects start from a position of proposing to change the way a business operates:

- Changing how processes and/or policies operate and the way people work.
- Changing the culture, the 'way we do things around here'.
- Impacting customers, employees and many other organizational stakeholders.

All projects have the potential to change the business in which they are being delivered; however, most have another driver:

- Developing a new project.
- Installing a new IT system.
- Building a new facility.

Poorly delivered business change, results in changes not being sustained, business benefits not being delivered (or sustainably delivered) and the potential for future business change to be unsuccessful. Organizations cannot afford to do this.

Organizations do not have limitless resources. They cannot do everything they want to do, and so now a more formal approach to these types of projects is typically being used. This can reap benefits for both the organization and the people within it. Figure 8-1 shows how a typical Business Change Project Roadmap uses stage gates to ensure that the right project is selected and then delivered successfully and sustainably.

- This particular roadmap does not consider any potential business change as a 'project' until it has formal approval within the organization. Until this point it is either an 'area of focus' or an 'idea'.
- A further stage gate ensures that no delivery (either design or implementation of a business change) commences until a robust delivery plan is in place and approved by the sponsor.

This type of project roadmap is inevitably 'front-end loaded', recognizing that early definition work does not generally impact people and allows for 'safe' review of a breadth of solutions to a perceived business issue. Additionally it filters out inappropriate changes earlier than usual, saving time, money and negative impact on those affected by such a change.

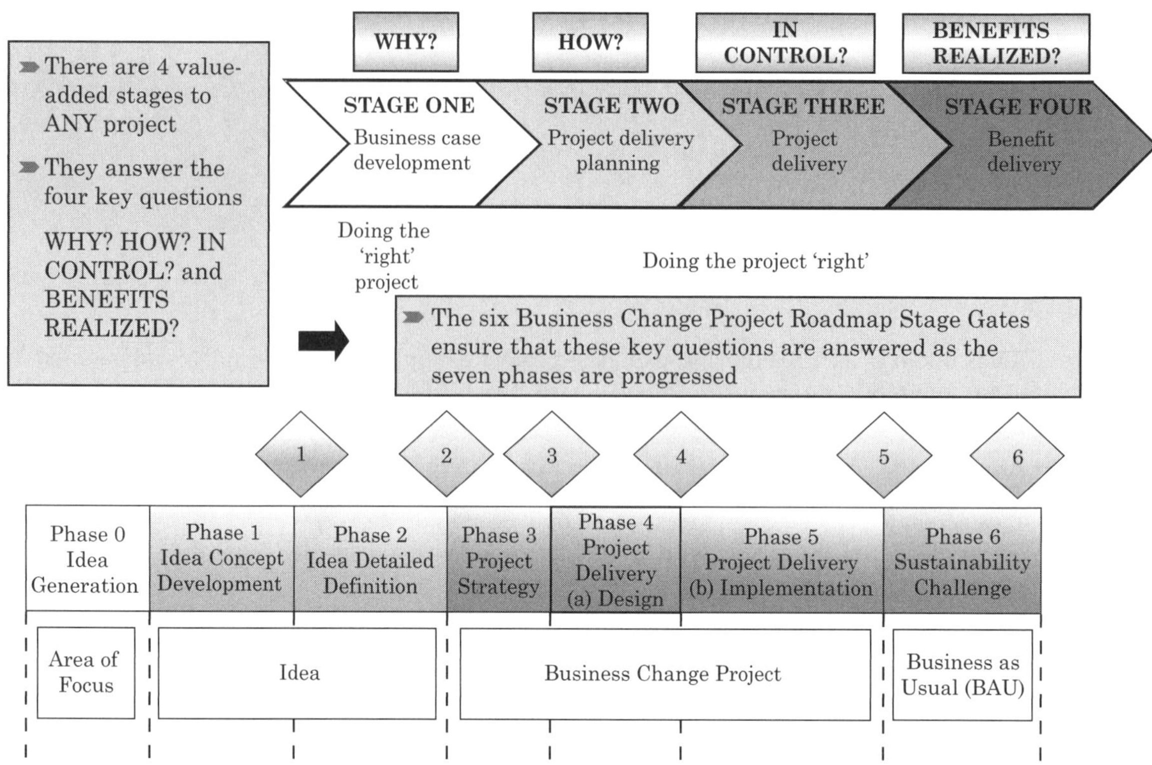

Figure 8-1 The generic project life-cycle and Business Change Project Roadmap

Alongside the project management toolkit, a Project Manager in a business change project needs to use a change management toolkit. This is a set of tools to support the identification, management, delivery and measurement of business change. Examples of such tools are shown in Table 8-1. This list is not exhaustive and those involved in business change projects are advised to develop a full change toolkit.

Within this case study project many of these tools were used to understand the problems which generated the 'idea' and then others were used to develop an appropriately designed solution.

2. Project outcome

This was a successful project. The reasons for success were analysed and are summarized in Figure 8-2.

It is worth noting that two of the five root causes represent the 'hard' side of project management and the other three the 'soft' side (see Chapter 2). This is an indication of the impact of 'human side' of business change projects.

Table 8-1 Example Change Toolkit

Change tool	Example use
Process map	Brainstorms all elements of a process (activities, decisions, documentation, etc.), linking them in readiness for a value-add (VA) review. Is also developed using data from observation of the process
VA review	The systematic evaluation of a process, identification of value-add and waste, then use of this evaluation to select appropriate opportunities for process improvement
IPO diagram	Assesses the required outputs from an activity and therefore the required inputs with appropriate process to achieve these outputs
Affinity diagram	A tool to gather and group ideas via brainstorming and then categorization
Five whys analysis	A root cause analysis technique (ask why 5 times)
Force field analysis	Assesses forces for and against a specific change
Time-value map	Analyzes the time taken to complete activities and determines if that time was value add or not
SWOT analysis	Assesses the strengths, weaknesses, opportunities and threats of a situation and then develops an action plan based on the analysis
As is/to be analysis	Maps the current process against the desired future process and reviews the gaps
Radar chart	Visually shows the size of gaps between current and desired performance against a set of defined performance criteria
Control chart	Monitors a process via specific performance criteria and studies its variation in order to define the course of the variation

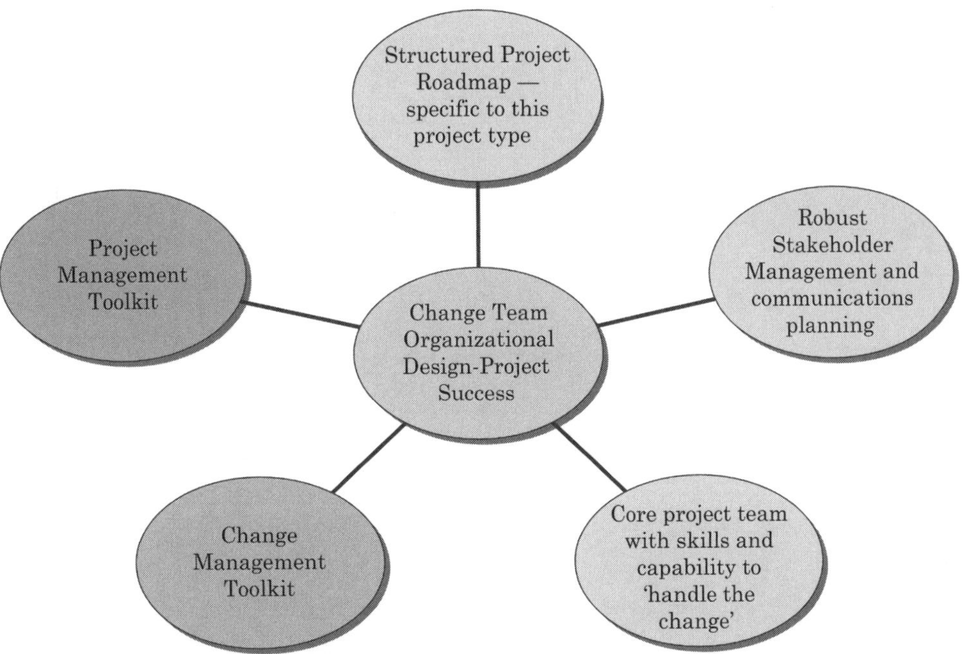

Figure 8-2 Root cause of project success

3. Actual tool use

The tools and methodology in *Project Management Toolkit* were used 'live' on the project and supported a successful outcome.

As a result substantial amounts of data were collected on the use of the tools. These are shared via use of the 'Lessons learnt' sections at the end of each of the four project stages and summarized at the end of this case study chapter (see page 281).

Situation

A change Project Team, responsible for supporting an organization in identifying and implementing changes to business processes (within the facilities management division of the business), has recently redeveloped the business process that they use (Figure 8-3). This new change process better supported selection of the right projects and then in doing them right!

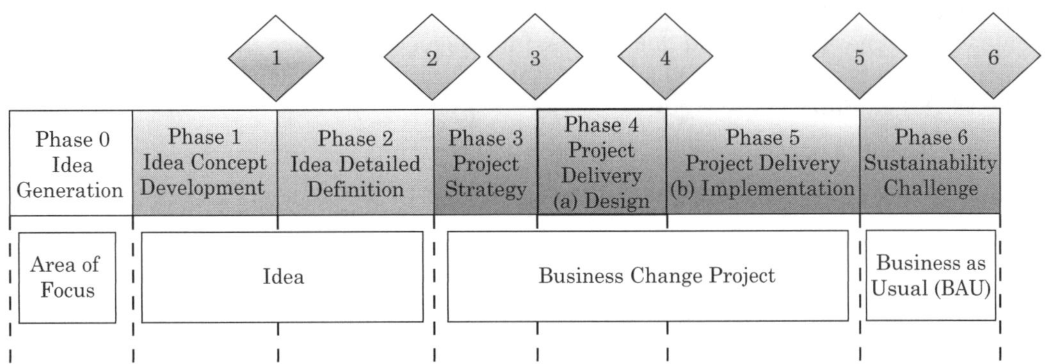

Figure 8-3 A Business Change Project Roadmap

- A new seven-phase process was developed. Five of these phases were either completely new or substantially changed:
  - Phases 0, 1 and 2 occurred before an 'area of focus' or 'idea' became fully approved and ensured that it was adequately investigated and articulated before business change was delivered. By the second stage gate the organization had a robust rationale as to *Why* it was doing the project.
  - Phase 3 introduced a planning phase following project approval and developed *How* the project was to be delivered.
  - Phases 4 and 5 cover the actual delivery and introduced further tools to ensure it was *In control.*
  - Phase 6 introduced a sustainability checking phase to check that the change continues to deliver the specified business benefits — *Benefits realized.*
- The process introduced six key decision points which were used as stage gates. This delayed, stopped or allowed the project to continue to the next phase depending on the robustness of the work and/or the appropriateness of the change in terms of aligned business benefits.

In implementing this change in their ways of working (WoW), further issues were highlighted, which, if not solved, would impact the sustainable benefits of the change. These were categorized as:

- Organizational issues — the way the team were organized and the roles they had.
- Business process issues — integrating the new change process within their existing WoW.
- Cultural — the values and behaviours within the team.
- Performance issues — the comparison of delivery performance vs external benchmarks.

A review of the impact of these issues confirmed that the project work at the heart of their 'business' could be delivered even more effectively if these issues were resolved. In addition, if the step change caused by the introduction of the new change process could be sustained then further improvements could be made through further change. This improvement would have the potential to optimise project delivery performance to meet a 'best in class' benchmark.

Figure 8-4 summarizes the situation and shows that two aligning 'ideas' were identified that would deal with the highest priority issues within the Change Team.

This case study follows the organizational design 'idea' through each of the four project stages and concludes after the first sustainability check, three months after the project had been fully delivered.

Inevitably there are links and dependencies between the two 'ideas' shown in Figure 8-4. Although both eventually became approved and then successfully delivered projects the Change Process Improvement Project is only referenced in so far as is necessary for the development of the Organizational Design Project case study.

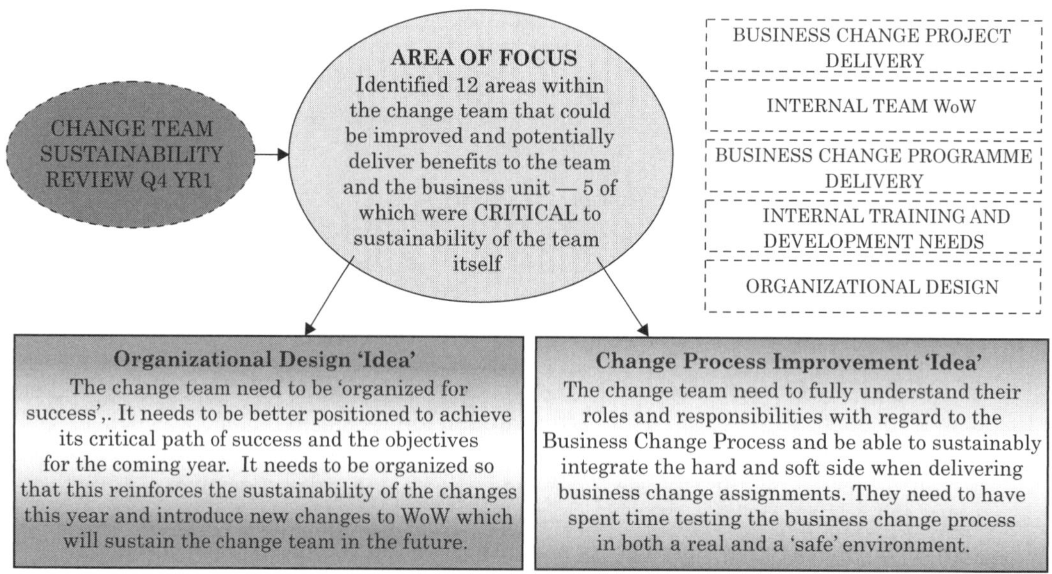

Figure 8-4 The reason for starting this project

The following sections illustrate the process which the Change Team Leader (Paul Smith) took to successfully complete the Change Team Organizational Design Project, assisted by his team of Portfolio Managers (Albert, Bill and Chris) and an External Consultant (Trish).

Stage One — Business Case Development

Once the potential project had been identified, the first phase was to define the 'idea' sufficiently in order to make a go/no go decision: to stop further work or to build a robust business case.

In the case of this change project the business case documentation was an organizational pre-requisite linked to the new change project process that the team had introduced.

Early stage problem definition

Within the new change process the initial Phase 1 work was done to address stage gate 1. If this 'idea' was implemented would the benefits align to the current business priorities? In order to answer this question two analyses were conducted: root cause analysis and development of a Simple Benefits Hierarchy.

- Root cause analysis is performed so that the eventual solution addresses the root cause rather than just the symptoms (Table 8-2).

Table 8-2 Symptoms and root causes

Symptom	Root cause	Proposed solution	Project
Project delivery cycle time exceeds benchmark	There are no internal project delivery benchmarks	Support project delivery through guidance, training, measures, tools and role development	**Organizational Design**
	Individuals have poor estimating skills when setting up project plans	Internal training and development — people and change process	Change Improvement Process
There are no consequences for lack of delivery performance	Inappropriate organizational structure and performance management processes	Improve Team WoW	**Organizational Design**
The portfolio of projects is not being actively managed	Inappropriate organizational structure and confused roles	Support portfolio delivery through measures, tools and role development	**Organizational Design**
Project Management skill gap	The team have a varied background and some are not formally trained in project management	Internal training and development — people and change process	Change Process Improvement
Change Management skill gap	The team have a varied background and some are not formally trained in change management	Internal training and development — people and change process	Change Process Improvement

(continued)

Table 8-2 (Continued)

Symptom	Root cause	Proposed solution	Project
The portfolio of projects exceeds the resource levels within the team or the business	Inappropriate business processes	Support project and portfolio management through measures, tools and role development	**Organizational Design**
The team members don't know what each other is doing	Inappropriate business processes and team WoW	Improve Team WoW	**Organizational Design**
Poor time and resource level estimation	There are no internal project delivery benchmarks	Internal training and development — people and change process	Change Process Improvement

A number of these analyses were conducted until the full set of high-level 'symptoms' had been reviewed. Figure 8-5 is an example of one of these analyses which started with the question 'Why does delivery cycle time exceed benchmarks?' Note that other symptoms come out as causes or effects.

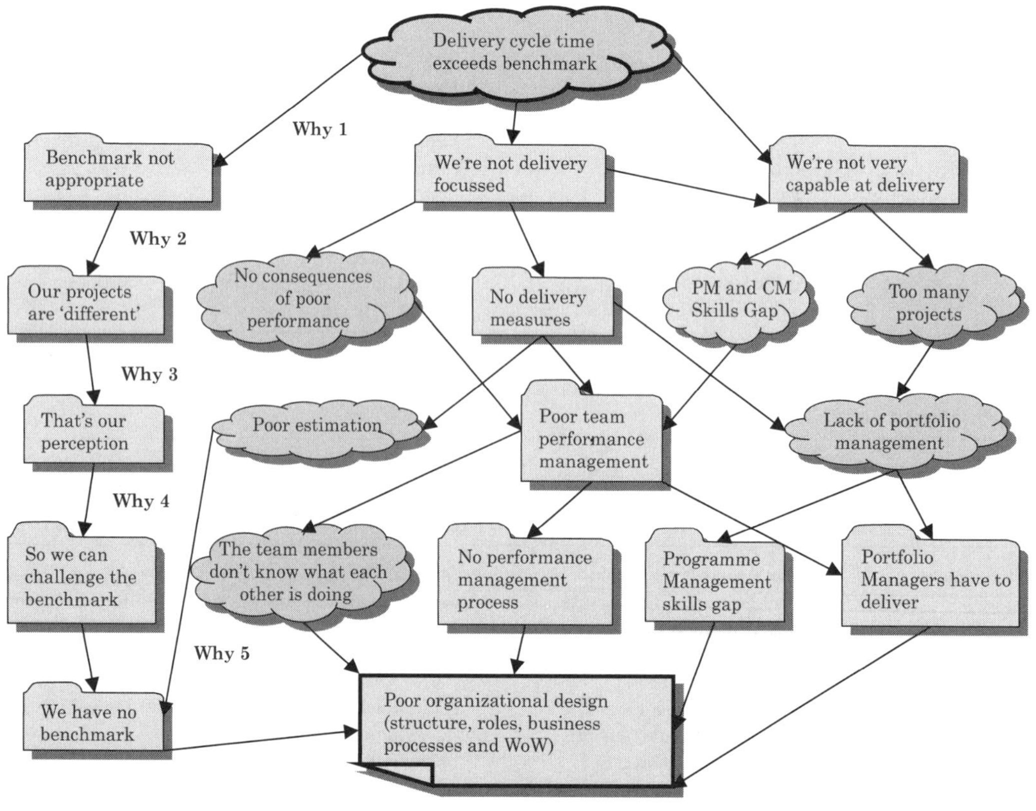

Figure 8-5 Root cause analysis

The root causes identified by this process confirmed the original 'idea': the Change Team needed to redesign itself. The level of redesign was not confirmed at this stage of the 'idea' development but was assumed to cover organizational structure, roles, WoW and business processes due to the results of the root cause analysis.

The Simple Benefits Hierarchy was then compiled and demonstrated that there was clear alignment between the benefits from implementation of this 'idea' and the needs of the business unit (Figure 8-6).

At this stage no actual design of the change itself had been completed, although a robust 'situational analysis' was completed which tested the feasibility of the main scope. The Change Team Leader conducted the majority of this Phase 1 work in liaison with his small management team. Specific aspects of team and project performance were discussed with the whole Change Team but only in the context of checking the sustainability of the previous year's change (the introduction of the Business Change Project Roadmap).

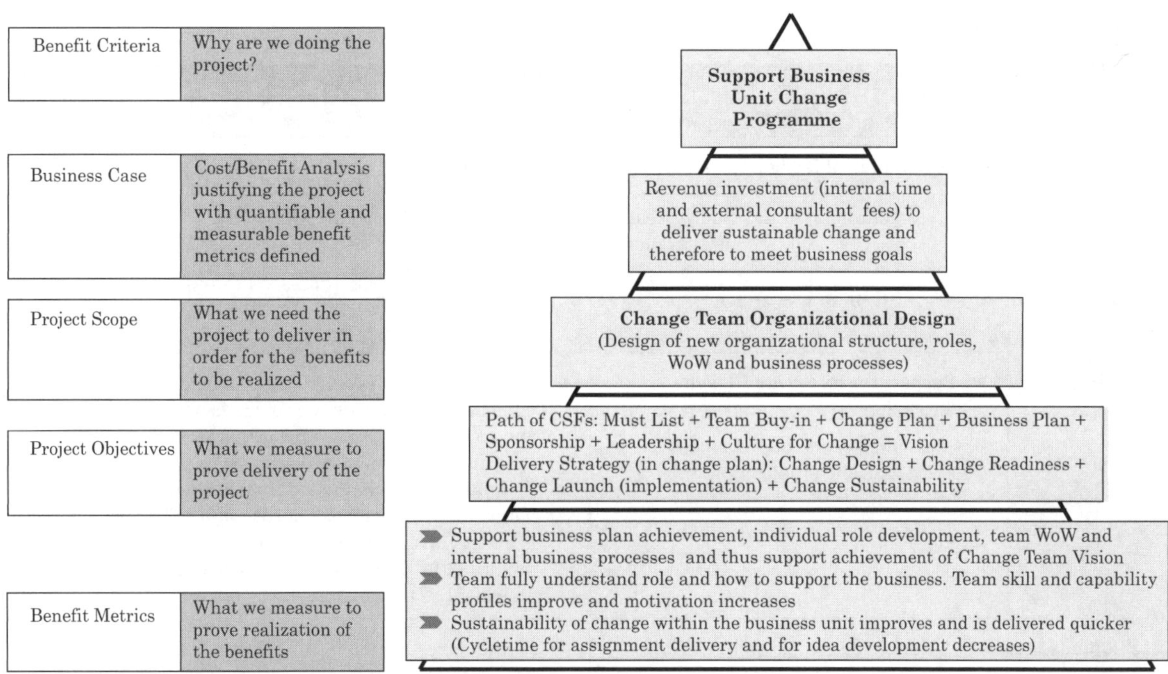

Figure 8-6 A Simple Benefits Hierarchy — Change Team Organizational Design

Business case development

The steering group within the business approved the 'idea' as presented by the hierarchy (Figure 8-6) and work on Phase 2 was then able to quickly progress.

This next phase required additional data and some conceptual design but only sufficient to develop a robust business case for organizational approval (stage gate 2):

- How much change is really required and how much resource will this need for successful delivery?
- How large is the impact on the business in the short term?
- What are the benefits for the business in the medium to long term?

The Detailed Benefits Hierarchy (Table 8-3) was used as the base document for this phase and helped to develop the work to an appropriate level. Additionally it reminded the team if they were slipping into the design phase, a typical problem seen in business change projects. The team remained the Change Team Leader (Paul Smith) with support from his management team (Portfolio Managers) although additional support was brought in using an External Consultant (Trish Roberts) who had supported the team in the development of the new Business Change Project Roadmap.

Table 8-3 Detailed Benefits Hierarchy

***Project Management Toolkit* — Detailed Benefits Hierarchy**		
Project: Change Team Organizational Design		**Date:** Month 2 Year 2
The reorganization of the Change Team, so that it is better positioned to support the business unit in which it operates		
Level in hierarchy	**Item in hierarchy**	**Comment**
1 Benefit criteria	➤ Support Business Unit Change Programme Business unit service improvement (effectiveness and efficiency)	➤ The business unit is a division of a large manufacturing company. Their role is to provide facilities management and so they are seen as an internal service company with internal customers only ➤ To continue to meet the external benchmark within the facility management sector. The business unit is under constant pressure to find service improvement opportunities and then quickly deliver this for the overall benefit of the company
2 Business case	➤ Revenue investment (internal time and external consultant fees) to deliver sustainable change to meet business goals Minimal expenditure for significant and tangible benefits	➤ Further investigation of the forecast improvement projects within the business has shown that there are significant financial and non-financial benefits which the business unit can deliver to the company ➤ The relatively small revenue expenditure required to implement this project is easily outweighed by these benefits

(continued)

Table 8-3 (Continued)

Project Management Toolkit — Detailed Benefits Hierarchy		
Project: Change Team Organizational Design		**Date:** Month 2 Year 2
Level in hierarchy	**Item in hierarchy**	**Comment**
3 Project scope	➤ Change Team Organizational Design ➤ (Design of new organizational structure, roles, WoW and business processes)	➤ See Figure 8-7 which articulates the levels of organizational design which this project needs to address ➤ Additional data review confirms the original scope in terms of categories but expands the required volume of scope to improve all WoW (business processes, team location and behaviours)
4 Project objectives	➤ Path of CSFs: Must List + Team Buy-in + Change Plan + Business Plan + Sponsorship + Leadership + Culture for Change = Vision ➤ Delivery Strategy (in change plan): Change Design + Change Readiness + Change Launch (implementation) + Change Sustainability	➤ See Figure 8-8 for the completed path of success for this project
5 Benefit metrics	➤ Support business plan achievement, individual role development, team WoW and internal business processes and thus support achievement of Change Team Vision ➤ Team fully understand their role and how to support the business. Team skill and capability profiles improve and motivation increases ➤ Sustainability of change within the business unit improves and is delivered quicker (cycletime for assignment delivery and for idea development decreases)	➤ See Table 8-4 for the complete Benefits Specification Table for this project

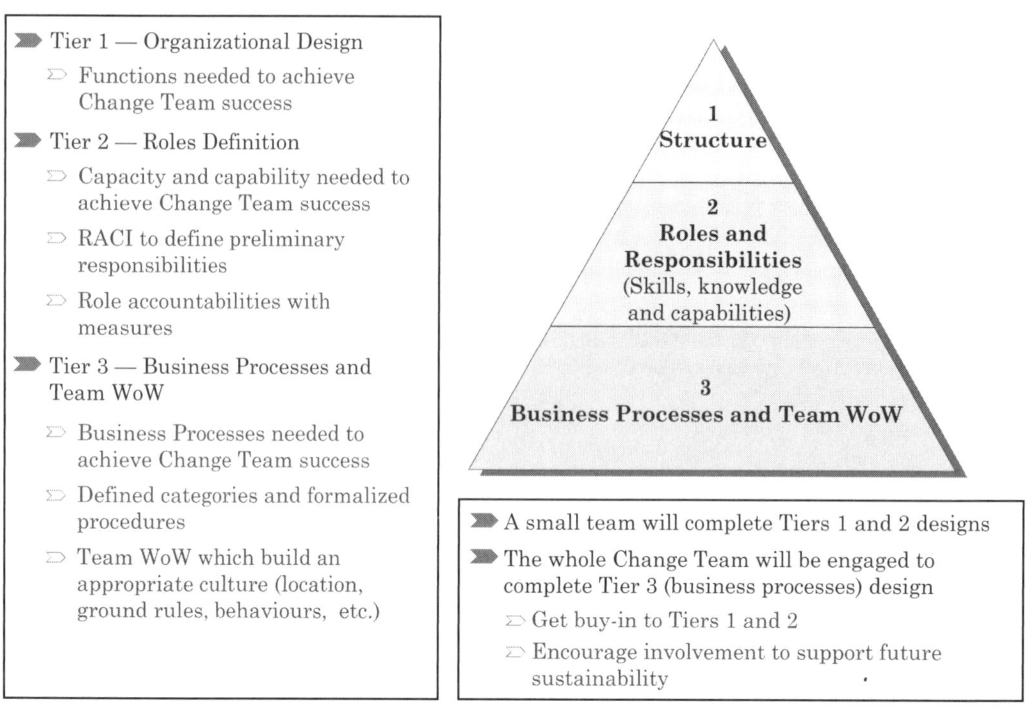

Tier 1 — Organizational Design
Functions needed to achieve Change Team success
Tier 2 — Roles Definition
Capacity and capability needed to achieve Change Team success
RACI to define preliminary responsibilities
Role accountabilities with measures
Tier 3 — Business Processes and Team WoW
Business Processes needed to achieve Change Team success
Defined categories and formalized procedures
Team WoW which build an appropriate culture (location, ground rules, behaviours, etc.)
1
Structure
2
Roles and Responsibilities
(Skills, knowledge and capabilities)
3
Business Processes and Team WoW
A small team will complete Tiers 1 and 2 designs
The whole Change Team will be engaged to complete Tier 3 (business processes) design
Get buy-in to Tiers 1 and 2
Encourage involvement to support future sustainability

Figure 8-7 Project scope

The development work during this phase involved:

- Mapping the current situation (formal and informal roles and structures).
- Reviewing the successes of the previous year and also the challenges for the next 1–3 years.
- Collecting team and project performance data.
- Conducting a full SWOT analysis (SWOT definition in Table 8-1).
- Completing a root cause analysis.
- Developing a vision of a successful change.
- Developing the scope definition (Figure 8-7).

Based on the preliminary work it was difficult for the Project Team NOT to start to consider what the new structure would look like. Trish, the external consultant, encouraged the team to develop a 'vision' in the form of a 'must list' but cautioned them against performing too much detailed design at this stage, when the 'idea' had not yet been approved as a project. This 'must list' became a crucial document throughout the project as it articulated the vision of a successful project in quite detailed terms:

- The processes which the Change Team would need to operate effectively.
- The skills needed within the team.
- How others in the business would perceive the team.
- How the team would perform — in terms of team performance and project performance.
- The outputs from the team in terms of deliverables to the business.

CSF 1 AN APPROPRIATE ORGANIZATIONAL DESIGN	CSF 2 AN ENGAGED CHANGE TEAM 'READY FOR CHANGE'	CSF 3 ROBUST CHANGE DELIVERY PLAN
Design when implemented meets the requirements as articulated by the 'must list'	The Change Team buy-in to the plan, design and implementation process. There is a culture of change readiness which informs the change plan	There is a robust change plan detailing the changes and the process. It is delivered successfully
CSF 4 SPONSOR SUPPORT	**CSF 5 CHANGE TEAM BUSINESS PLAN**	**CSF 6 LEADERSHIP**
The Business Leader supports the changes which align to his business plan and the CSF he needs to achieve it	The change team must deliver its in-year plan (business as usual) whilst aligning with the proposed internal organizational changes	The management team within the change team demonstrate support of this change 'no matter what'

PROJECT VISION
The change team are 'organized for success' and can deliver sustainable and agile change within the business

Figure 8-8 Path of success — Change Team Organizational Design

The team were surprised by the completed path of success (Figure 8-8) as they had assumed that the main focus would be on the structural design — deciding who would report to whom, etc. However, the CSFs reiterated the importance of (a) articulating the basis for design and (b) planning the change before delivery.

As a result of the idea, development work on a business case was collated (Table 8-5)

As a final check Paul, the Change Team Leader, asked Trish, the External Consultant to perform a quick 'health check' prior to submitting the business case for formal approval (Table 8-6). The 'health check' confirmed that the 'idea' was sufficiently developed for the business case approval stage gate.

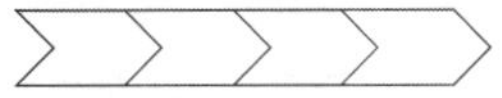

Table 8-4 Benefits Specification Table

Project Management Toolkit — Benefits Specification Table					
Project: Change Team Organizational Design			**Date:** Month 2 Year 2		
Potential benefit	**Benefit metric**	**Benefit metric baseline**	**Accountability**	**Benefit metric target**	**Area of activity**
What the project will enable the business to deliver	Characteristic to be measured	Current level of performance	Accountable for delivery of the benefit to target	Performance to achieve overall benefits	The project scope that will enable this benefit to be delivered
More successful change in year	Change Team utilization (% of time on core projects)	60%	Change Team Manager	80%	Structure, Roles and Business Processes
	% of projects progress to plan	50%	Portfolio Managers	90%	
	Portfolio speed (projects/year)	5		10	
Improved project delivery performance (faster and more effective)	Cycle time — Business Case approval (0–2)	16 weeks	Change Team Manager	8 weeks	Role Development and Business Processes
	Cycle time — approved plan (Phase 3)	—		4 weeks	
	Cycle time — delivery (4 and 5)	35 weeks		22 weeks	
Improved Sustainability of Change	% of benefits delivered to plan	50%	Change Team Manager	90%	Role Development and Business Processes

Table 8-5 The Business Case Tool

Project Management Toolkit — Business Case Tool			
Project: Change Team Organizational Design		**Date:** Month 2 Year 2	
Business case developed by	Paul Smith (Change Team Leader)	**Date**	Month 2 Year 2
Project reference number	FM-138	**Business area**	The Change Team within the Facilities Management Unit
Project Manager	Paul Smith (Change Team Leader)	**Project sponsor**	John MacDonald (FM Business Unit Leader)
Business background	During year 1 the Change Team developed a more structured and formal business change project roadmap (Figure 8-3). Apart from the benefits realized in terms of improved project delivery, stopping inappropriate projects earlier and improved use of Change Team resources, the new roadmap showed there were still improvements to be made if the team were to deliver to external benchmark levels		

(continued)

Table 8-5 (Continued)

<table>
<tr><th colspan="4">Project Management Toolkit — Business Case Tool</th></tr>
<tr><td colspan="2">Project: Change Team Organizational Design</td><td colspan="2">Date: Month 2 Year 2</td></tr>
<tr><td>Project description</td><td colspan="3">Once this project has been successfully completed the Change Team within FM will be 'organized for success'
➤ It will be better positioned to meet its critical path of success and the objectives for year 2
➤ It will reinforce the sustainability of the changes in year 1 and introduce new changes to WoW which will sustain the team in the future
➤ It will have an appropriate structure, internal capability and business processes to support transformation within the FM Business Unit</td></tr>
<tr><td>Delivery analysis</td><td colspan="3">Methodology — the Business Change Project Roadmap (Figure 8.3) will be followed for the whole project. The design methodology would use a standard organizational design process
Resources — part-time resources made up of Paul Smith and his three Portfolio Managers (Albert, Bill and Chris) plus support on a 1 day/week basis from Trish Roberts, the External Consultant, who has been working with the Change Team in year 1
Dependencies — this is linked to the current use of the Business Change Project Roadmap, the Change Process Improvement Project (also going through formal approval at this time) and the changing project priorities as linked to the FM Business Unit Plans
Risks — Currently the risks are assessed as 'amber' and this is mainly due to the extent of change being delivered within one small team plus the usual 'amber' risks associated with changing roles and structures. It is strongly believed that these can be appropriately mitigated
Schedule — To be delivered with the standard benchmark cycle time for a medium size/medium complex project. Implementation in 2 phases: structure and roles by end of month 5, business processes and WoW by end of month 7. To be fully sustained by end of year 2</td></tr>
<tr><td>Business change analysis</td><td colspan="3">The project will impact current change assignments being delivered in year 2; however the short-term impact should be outweighed by the improved speed of delivery of projects in the medium term
It is likely that the project will impact the flow of work through the Change Team. In order to gain some speed the portfolio may need to be further reduced so at any one time there are fewer projects being delivered but the same level of activity will occur 'in year' overall</td></tr>
<tr><td>Value-add analysis</td><td colspan="3">To continue to meet the external benchmark within the facility management sector, the business unit is under constant pressure to find service improvement opportunities and then quickly deliver these for the overall benefit of the company</td></tr>
<tr><td>Impact of NOT doing the project</td><td colspan="3">If the project is not delivered then it is likely that delivery performance will remain as. This may then have an adverse impact on:
➤ The sustainability of the Business Change Project Roadmap and wider introduction into smaller change projects delivered within the business
➤ The delivery of the service improvements as needed by the FM Business Unit</td></tr>
<tr><td>Project approved
(Value-add or not?)</td><td>Yes/~~No~~</td><td>Name of approver and date</td><td>FM Business Unit Change Project Steering Team — Approved Month 2 Year 2</td></tr>
</table>

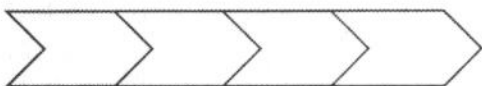

Table 8-6 The 'Why?' Checklist

***Project Management Toolkit* — The 'Why?' Checklist**	
Project: Change Team Organizational Design	**Project Manager:** Paul Smith
Date: Month 2 Year 2	**Page:** 1 of 2
Sponsorship	
Who is the sponsor? (The person who is accountable for the delivery of the business benefits) Business Unit Manager — John MacDonald **Has the sponsor developed an external communication plan? (How the sponsor will communicate with all stakeholders in the business)** No — however in discussions with the Project Manager the sponsor has agreed to update his leadership team so that they understand the challenges on the Change Team over the coming months. The agreed key message is that these short-term challenges will reap major benefits for improvements in all parts of the business in the medium and longer term	
Business benefits	
Has a business case been developed? Yes — a formal business case document has been collated (Table 8-5) and is due to be submitted to the FM Project Steering Team that was set up as a result of the introduction of the new Business Change Project Roadmap. **Have all benefits been identified? (Why is the project being done?)** Yes — see Table 8-4 — although understanding the size of some of these benefits will require tracking to get a baseline **Who is the customer? (Identify all stakeholders in the business including the customer)** A change within a Change Team always causes confusion as regards identification of the customer (the person or group who will benefit from the project). In some respects the whole Change Team, the sponsor, the business unit leadership team and any customers of the business unit itself all represent customer groups. The Change Team Manager is aware of this complexity and although a formal stakeholder analysis has not yet been completed it is not 'out of control' **How will benefits be tracked? (Have they been adequately defined?)** Table 8-4 will be converted into a tracking tool	
Business change	
Will this project change the way people do business? (Will people need to work differently?) Yes — the Change Team will go through a major change as their roles will inevitably be altered **Is the business ready for this project? (Are training needs identified or other organizational changes needed?)** No — the business doesn't understand the impact that this change will have on the support to their change projects. However, this is not an issue yet because the scale and breadth of business impact is not fully articulated and cannot be until the detailed design is completed. The Change Team are also not aware of the proposed changes but again it is not appropriate to communicate this until the project is formally approved and some design work is under way. We need to have clear messages to communicate so it's RIGHT that the business is not ready However, the business does need this project and the sponsor has made assurances that he will support any subsequent business readiness issues	

(continued)

Table 8-6 (Continued)

Project Management Toolkit — The 'Why?' Checklist	
Project: Change Team Organizational Design	**Project Manager:** Paul Smith
Date: Month 2 Year 2	**Page:** 2 of 2
Scope definition	
Has the scope been defined? (What level of feasibility work has been done?) A good preliminary scope has been developed based on a review of data, development of a path of success and an understanding of the generic scope for organizational design projects **Have the benefit enablers been defined? (Are you sure that the project will enable the benefits to be delivered when the project is complete?)** There is a clear link between agile project delivery and service improvement business benefit realization **Have all alternatives been investigated? (Which may include *not* needing the project)** Yes — the analysis demonstrates that both the process and the organizational design require change and the two connected projects are required **Have the project success criteria been defined and prioritized?** Yes — path of success, project vision and a set of benefits describe success. In the next stage sustainability planning need to be reviewed and this may support definition of other associated changes in the wider business unit as well as the Change Team	
Stage One decision	
Should the project be progressed further? (Is the business case robust enough for detailed planning to commence?) Yes — this project has a robust business case with clearly aligned benefits for the business unit	

End of Stage One situation

The project was able to progress through Stage One very quickly and so went from 'idea' to approval in 5 weeks. This matched a 'best practice' benchmark for a change project of this breadth.

Approval was gained to deliver the project and a high degree of confidence was expressed due to the quality of the Stage One work.

Lessons learnt

A brief after-action review (AAR) highlighted the following key lessons learnt:

- Resist the temptation to 'dive into the design'. Design work is more robust if based on a clear rationale backed up by objective data.
- Do evaluate the 'symptoms' and ensure that the 'idea' will solve the root cause.
- Minimize the level of communication of the 'idea'. Over communicating poor or draft information can be as harmful in a change project as communicating nothing at all. Develop some key messages and communicate with the right stakeholders. Bear in mind how people naturally react to change: they resist it!. Remember that this is only an 'idea' and communicate appropriately.
- Ensure that the early definition work is succinctly completed based on objective data.

Stage Two — Project Delivery Planning

Following project approval the small team was formalized and began to plan how this organizational change would be designed and implemented so that the benefits could be realized as soon as possible. Figure 8-9 shows the formal links between the Project Team, sponsor and key stakeholders:

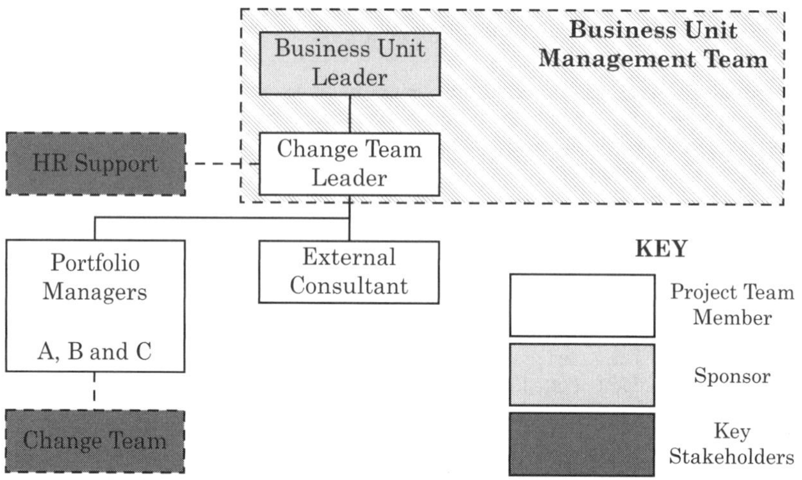

Figure 8-9 Project organization

The Project Manager and the three Portfolio Managers were only able to spend 50% of their time on the project; so the Project Manager (Paul Smith) asked the consultant (Trish Roberts) to take a lead in this stage. It was clear that the whole team wanted to model 'practice what you preach' behaviour. They wanted a robust delivery plan in place before any delivery began and also before any formal communications to the wider team.

Trish used the 'How?' Checklist as a guide throughout the 4-week period and converted it into a short-term action plan (Table 8-7). Based on Table 8-7 the planning commenced — initially with a definition of the overall project strategy and then with the development of the detailed planning deliverables as highlighted.

Table 8-7 Planning action plan

'How?' Checklist	Document to generate	Dependencies or issues
1. Any changes which will impact the business case?	➤ None	➤ Cross-check the table of CSFs with the scope in the business plan ➤ Cross-check the Benefits Specification Table with the one in the business plan
2. Sponsor situation?	➤ Communications plan	➤ Paul Smith to agree key messages with sponsor based on stakeholder analysis

(continued)

Table 8-7 (Continued)

'How?' Checklist	Document to generate	Dependencies or issues
3. Benefit realization planned?	➤ Benefits Specification Table Rev 2 ➤ Benefits Tracking Tool	➤ Check that these can be measured ➤ Check that CSF delivery will enable these benefits ➤ Review likely projects timetable so that the benefits realization can be forecast and specific milestones set
4. Business changes planned?	➤ Stakeholder Management Plan ➤ Draft Sustainability Plan	➤ Based on draft plan from Stage One ➤ The sustainability plan needs to identify 'early warning' measures that would show that the benefits are or are not going to be delivered
5. Scope defined?	➤ Table of CSFs	➤ Based on CSFs identified in Stage One
6. Project process defined?	➤ Organizational design process defined ➤ Phasing of project defined vs stage gates	➤ Use base organizational design process ➤ Base phasing on scope and need to get this delivered ASAP
7. Finance sorted?	➤ Resourcing plan	➤ Link to phasing of project delivery
8. Risks managed?	➤ Risk Table and Matrix Rev 2	➤ Ongoing review of risks and development of mitigation plans
9. Project Team defined?	➤ Project Organogram ➤ Project RACI Chart	➤ Link to phasing of project delivery
10. External support?	➤ External Consultant work scope and plan	➤ Link to gaps in Project Team capability and experience
11. Review plan in place?	➤ Milestone review plan	Link to phasing of project delivery

Project strategy

The basic strategy for the project would generally follow a typical change project methodology:

- *Change Design* — collecting all the appropriate data, completing the diagnosis and developing a detailed design of how the organization should operate — including organizational structure, roles and WoW.
- *Change Readiness* — assessing the culture for change in the team and individuals. Reviewing capabilities and "matches" with the new organization. Communication to develop an appropriate level of involvement and excitement for the change.
- *Change launch* — phased implementation of the design: communicating the change, role selection and implementation of the changes to the organization, roles and WoW.

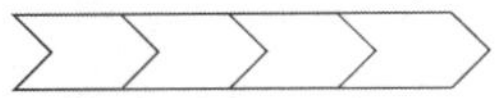

All the Project Team agreed that it was going to be difficult to continue to work together at such a high level and not expect the team to be curious about the reasons. Therefore, any overall delivery strategy must incorporate an early communication of the intent and scale of the change. However, for this to be effective it must include some significant information that allows the Change Team to ask themselves 'What does this mean to me?'. Without this data, team morale and trust will be impacted. To align with this need Trish recommended a 2-phase strategy based on the typical methodology discussed:

- Phase 1 — design and implement the organizational structure and roles.
- Phase 2 — design and implement the new business processes and team WoW.

This would have a number of benefits:

- Phase 1 could be very quickly designed, based on the extent of data already collected and analysed leading to:
  - An early communication of the structural change: Who reports to whom, what the new roles are and the process by which current team members would be placed in those roles.
  - A conceptual overview of the business processes and team WoW needed to support the above.
- Phase 2 would be the detailed design of the business processes and team WoW and involve the wider team:
  - Team participation in design supports a more effective engagement process which in itself supports the change being sustained.

Based on this, a change plan was pulled together in line with Figure 8-10 and this formed the bulk of the project delivery plan along with the documents mentioned in Table 8-7.

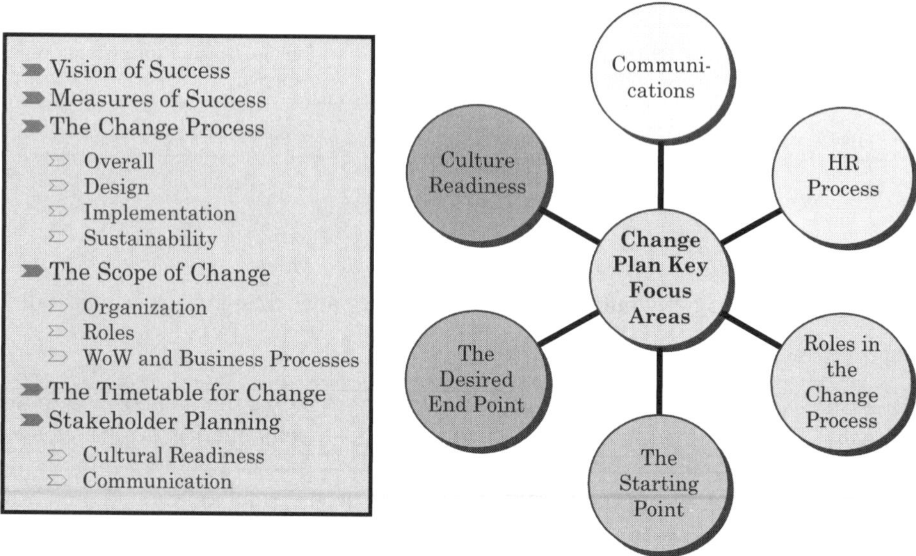

Figure 8-10 The Change Plan — Contents

This particular business change project is going to deliver an organizational design and there is a very specific design process to go through. This is shown in Figures 8-11 and 8-12.

In effect, the plan was that Phase 1 of this project would go through all the steps in Figure 8-11 in order to develop a 'base design' which could be approved prior to implementation.

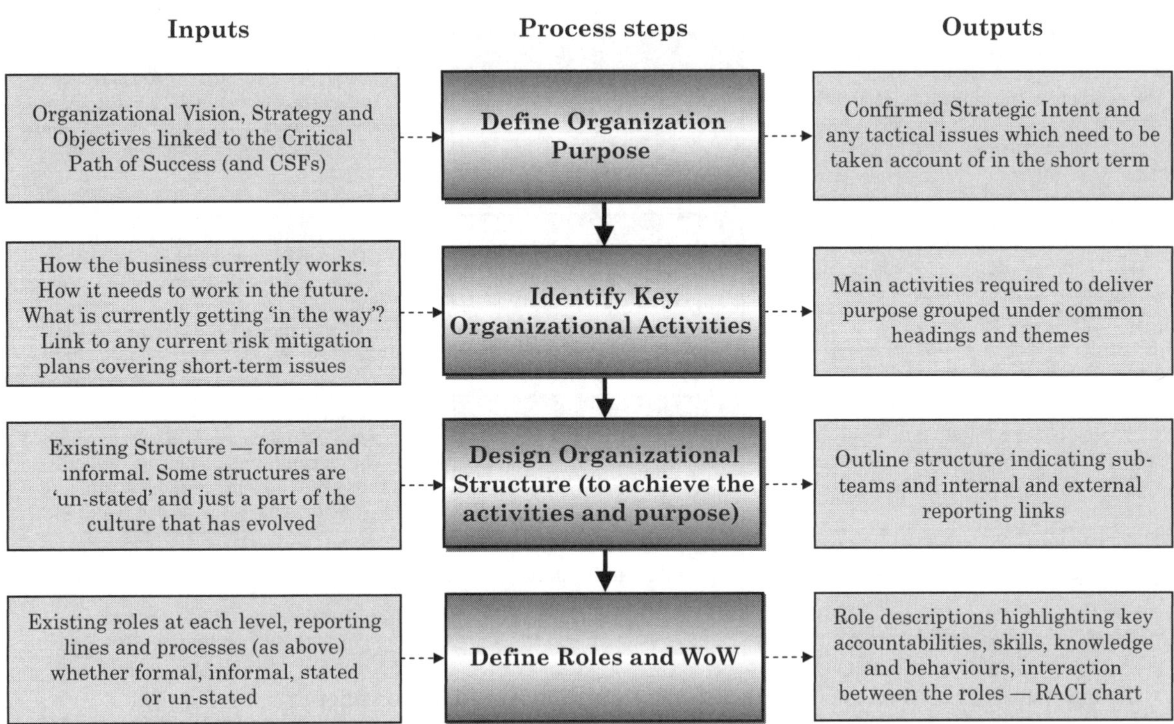
Inputs
Process steps
Outputs
Organizational Vision, Strategy and Objectives linked to the Critical Path of Success (and CSFs)
Define Organization Purpose
Confirmed Strategic Intent and any tactical issues which need to be taken account of in the short term
How the business currently works. How it needs to work in the future. What is currently getting 'in the way'? Link to any current risk mitigation plans covering short-term issues
Identify Key Organizational Activities
Main activities required to deliver purpose grouped under common headings and themes
Existing Structure — formal and informal. Some structures are 'un-stated' and just a part of the culture that has evolved
Design Organizational Structure (to achieve the activities and purpose)
Outline structure indicating sub-teams and internal and external reporting links
Existing roles at each level, reporting lines and processes (as above) whether formal, informal, stated or un-stated
Define Roles and WoW
Role descriptions highlighting key accountabilities, skills, knowledge and behaviours, interaction between the roles — RACI chart

Figure 8-11 Organizational design process (a) getting to a 'base design'

Although the Change Team had previously discussed the project and likely outcomes with HR (Human Resources Department) it is at this stage that the detail of the "base design" would be discussed with them. This allowed agreement on the appropriate method of getting people into the roles. If the roles are not significantly different then there may be a way of 'matching' roles to avoid lengthy selection processes. Based on the capabilities and experiences in the team it was expected that this would be the eventual HR process. All the Project Team recognized that this was a major schedule and people risk. If team members had to reapply for their jobs it would not only take time but would be highly disruptive and potentially demotivating for the team.

The second part of the Organizational Design (OD) process (Figure 8-12) would take Phase 1 to completion.

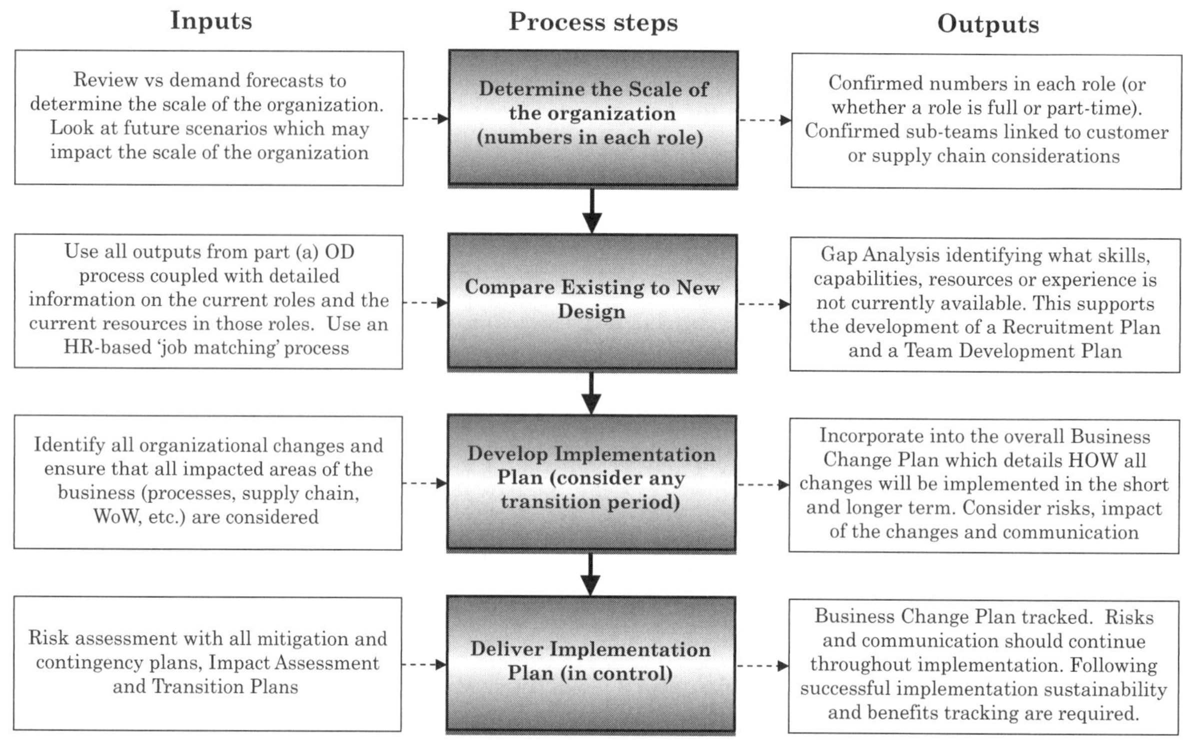

Figure 8-12 Organizational design process (b) from design to implementation

Throughout each step of the OD Process the following would be considered:

- Current and future organizational culture.
- The impact of the change on the people involved in the change.
- The impact of the change on the customers and other stakeholders.
- Stakeholder management via appropriate consultation and communication.

Following successful implementation the following would be incorporated into the sustainability plan:

- Handover of all project deliverables — ensuring the completion of all required project scope.
- Benefits tracking — ensuring that all the business reasons for the change are being delivered.
- Sustainability tracking — ensuring that the change is being sustained using appropriate sustainability measures, for example:
  - Organizational culture — what can be seen, how people behave, etc.
  - Customer feedback, etc.

Aspects of the project delivery plan

Once the overall strategy for the project was agreed, additional data collection commenced so that more detailed delivery planning could be done.

All the deliverables in Table 8-7 were developed; however, the areas of the plan which took the most time to complete were those supporting the 'softer' side:

- Stakeholder Management Plan — reviewing the impact of the change on individuals and groups and developing specific plans to engage them as much as possible and as necessary.
  - The Change Team were the major group of people impacted by this change on a personal level. As a result of this project their whole way of working, who they report to and their daily roles and responsibilities would change.
  - The current projects under delivery that were being managed by the Change Team all had current business users, customers and sponsors. The impact of change on the people within these projects (and potential impact on the benefits to be realized) needed careful review.
  - Significant time was spent on reviewing likely responses to change.
- Culture Review Plan — reviewing the current 'way we do business' and articulating the changes, ensuring that the project has considered these during the design phase.
  - A lot of this work was done as a part of assessing the benefits from the project and the potential sustainability of change.
  - In general the view was taken that a Change Team has a higher threshold for step change as it is usually they who are causing it for others.
  - Such assumptions were tested through observation as a part of the planning stage.

Both of these supported the development of the Communications Plan (Table 8-8), understanding the key messages which need to be consistently communicated at specific phases in the project.

Table 8-8 Project Communications Plan

Project Management Toolkit — Communication Plan		
Project: Change Team Organizational Design	**Date:** Month 3 Year 2	
This Communications Plan is specifically designed to support the design, implementation and sustainability of the Change Team Organizational Design Project		
Audiences		
Project sponsor: John MacDonald	**Communications sub-team:** Trish and Chris	**Project stakeholders:** Customers of the Change Team, project team members from FM Business Units
Project lead: Paul Smith	**Project customer:** The Change Team	
Objectives		
• To identify all audiences (internal and external) at the various project phases and to ensure their communication needs are met • Build broad project awareness for all stakeholders • Communicate project plans to all those impacted by the change (project customers). This must be timely and specific to their needs • Provide a stable, trustworthy and consistent source of information • Support the Change Team in cascading appropriate information to their Project Teams		

(continued)

Table 8-8 (Continued)

Project Management Toolkit — Communication Plan

Project: Change Team Organizational Design	**Date:** Month 3 Year 2

Measures of success

- Support the Change Team so that they are motivated and engaged in the project
- Model good communication skills and promote their development within the Change Team
- Encourage feedback from the Change Team
- Project AAR shows communication as a 'what went well'
- No project crisis which when root caused links back to poor communication
- Direct feedback from the Change Team on the appropriateness of the communications (various vehicles including surveys, informal, focus groups, etc.)

Project phase	Communication objectives	Broad communication activities
Phase 1 Design	➤ To contain any details of the change project until the design had sufficiently progressed (but not to hide the fact that the management team was working on this) ➤ To build awareness that change was coming	➤ AAR for previous change project ➤ Trish support presentation ➤ Introduction of the vision for the Change Team and the detail of the objectives for Year 2
1st Communications End of Phase 1 Design	➤ To inform the Change Team and the Business Unit Management Team of the project — scope, roles, timing, vision of success and current status ➤ To describe the new *Structure* and *Roles* ➤ To describe the process to launch the above	➤ Communications meeting with presentation, handout and Q & A session ➤ Presentation involves all Project Team
Phase 1 Implementation	➤ To inform the Change Team of individual changes where possible and to ensure that individuals are appropriately supported ➤ To inform the Change Team of project status and the future plans and targets	➤ Manager one-to-ones (using HR as appropriate) ➤ Regular one-to-one, buddy one-to-one and Project Team 'open door' ➤ Team surveys
Phase 2 Design	➤ Enabling appropriate team involvement in Tier 3 design (focused support effort) ➤ To inform the Change Team of project status, future plans, targets and their role in awareness communications ➤ Project Team model appropriate, motivated and supporting behaviours related to the changes ➤ Convey to the Change Team and the sponsor that business as usual (BAU) is critical and Project Team model BAU behaviours	➤ Weekly bulletin and focus session (15 minute weekly face to face around the bulletin board and incorporate a short Q & A session) ➤ Regular one-to-one, buddy one-to-one and Project Team 'open door' ➤ Team surveys ➤ Focus group sessions, as appropriate ➤ Use team meetings as appropriate

(continued)

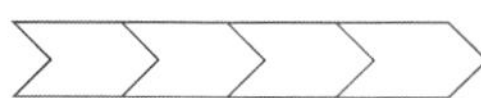

Table 8-8 (Continued)

Project Management Toolkit — Communication Plan		
Project: Change Team Organizational Design	**Date:** Month 3 Year 2	
Project phase	**Communication objectives**	**Broad communication activities**
Launch End of Phase 1 Implementation	➤ To make sure that the Change Team see that things are different ➤ To support the Change Team through this change (launch week) ➤ Project Team model appropriate, motivated and supporting behaviours, are visible and available ➤ To support the Change Team in cascading general awareness that change is coming for their teams and sponsors	➤ Office layout changes and posters (new team identities) ➤ Launch event — Team lunch and communication of key messages ➤ Project Team 'open door'
Post Launch Phase 2 Design and Implementation	➤ To support the Change Team through this change ➤ Project Team model appropriate, motivated and supporting behaviours, are visible and available ➤ To inform the Change Team of the status of the project (know when the project is finished) ➤ To support the Change Team in cascading specific changes as related to their projects to their teams and sponsors ➤ Convey to the Change Team and the sponsor that 'BAU' is critical and Project Team model 'BAU' behaviours	➤ Weekly bulletin and focus session (15 minute weekly face to face around the bulletin board) ➤ Helpdesk surgery sessions ➤ Workshop linked to Change Process Improvement Project ➤ End of project event ➤ Regular one-to-one support ➤ Team surveys ➤ Focus group sessions, as appropriate ➤ Use team meetings as appropriate
Project Completion Sustainability	➤ Convey to the Change Team and the sponsor that 'BAU' is critical and Project Team model 'BAU' behaviours ➤ To support the Change Team in reaching 'steady state'	➤ Helpdesk surgery sessions, as appropriate ➤ Regular one-to-one support ➤ Sustainability check ➤ Project AAR
Business as Usual (BAU)	➤ The Change Team Communications Plan is developed and managed by the new Communications Lead	➤ As outlined within the Communications Plan

During this time the majority of the 'harder' side of project delivery planning was completed by Trish working in partnership with one of the Portfolio Managers:

- Development of a work breakdown structure through use of the table of CSFs (Table 8-9).
- Use of the above to generate a detailed RACI Chart so that each aspect of the project could progress upon approval of the plan (Table 8-10).
- Assessment of risk though use of the path of success with the Risk Table and Matrix (see summary in Table 8–11).
- Additional work to compile resource plans, schedules and commence tracking benefit metrics in order to gain a 'baseline.'

Table 8-9 Table of Critical Success Factors

***Project Management Toolkit* — Table of Critical Success Factors**			
Project: Change Team Organizational Design		**Date:** Month 3 Year 2	
Critical path of success			
The team are 'organized for success' and can deliver sustainable and agile change within the business			
Critical success factors definition			
Scope area (CSF level 1)	**Objective tracking metric (CSF level 2)** ***Next level of scope to deliver CSF level 1***	**Critical milestone (CSF level 3)** ***Next level of scope to deliver CSF level 2***	**Priority (only HIGH to be noted)**
CSF 1 — An Appropriate Organizational Design *Design when implemented meets the requirements as articulated by the 'must list'*	Design basis	'Must list'	High
	Structural design	Organization design	
		Role design	
	Business process design	10 business processes designed	
	Team WoW	Office layout design	
		Team ground rules defined	
CSF 2 — An Engaged Change Team 'Ready for Change' *The Change Team buy-in to the changes and to the design and implementation process. There is a culture of change readiness which informs the change plan*	Cultural development	Define desired 'end state'	High
		Cultural readiness assessments	
	Communications	Launch communications	High
		Weekly bulletins	
		Manager one-to-ones	High
		Buddy system	
CSF 3 — Robust Change Delivery Plan *There is a robust change plan change process and it is delivered successfully needs to achieve*	Change process	Path of success	High
		OD process definition	
		HR process definition	
	Change delivery	Launch organization	High
		Appoint into roles	High
		Launch Business Processes	
		Change office layout	
		Launch team WoW	

(continued)

Table 8-9 (Continued)

Project Management Toolkit — Table of Critical Success Factors

Project: Change Team Organizational Design | **Date:** Month 3 Year 2

Scope area (CSF level 1)	Objective tracking metric (CSF level 2) *Next level of scope to deliver CSF level 1*	Critical milestone (CSF level 3) *Next level of scope to deliver CSF level 2*	Priority (only HIGH to be noted)
CSF 4 — Sponsor Support *The Business Leader supports the changes which align to his business plan and the CSFs he needs to achieve*	Contracting	Support of 'idea'	High
		Project approval	
	Communication	Monthly bulletins	
CSF 5 — Change Team Business Plan *The Change Team must deliver its in–year plan (BAU) whilst aligning with the proposed internal organizational changes*	Business as usual delivery	Track assignment KPIs	High
		Track resource issues	
		Track customer issues	
	Stakeholder management	Stakeholder Management Plan	High
		Communications Plan	
CSF 6 — Leadership *The management team within the Change Team demonstrate support of this change 'no matter what'*	Model appropriate behaviours	Team review and feedback sessions	High
	Robust project management	Progress tracking	High
		Regular risk reviews	

Table of CSFs

Trish and Albert reviewed the path of success (Figure 8-7) with the team in order to develop the table of CSFs shown in Table 8-9. They found the prioritization of each level 3 CSF difficult as there was so much dependency between them. In the end they decided to only note the *high* priorities. These tended to be the key area which the rest of the scope was dependent on. Typically these were to be completed in Phase 1 of the project. This table will be converted into the Milestone Tracking Tool when the project moves into delivery (page 270).

The accountability for delivery of each element of the scope as defined by the CSF table was articulated via the RACI Chart (Table 8-10).

RACI Chart

Bill worked with Trish to evaluate how much time each team member had and the most appropriate way to segment the various activities in order to develop the RACI Chart (Table 8-10). The RACI definitions were altered slightly to accommodate the particular project situation, for example some design was via team work involving all. However, where possible each activity has only one team member accountable and one responsible. Where multiple responsibilities were needed a more detailed RACI was produced, for example the business processes were planned at a more detailed level.

Table 8-10 Project RACI chart

Project Management Toolkit — RACI Chart

Project: Change Team Organizational Design	**Date:** Month 3 Year 2
R Responsible for actually doing the activity	C^R Active support in delivery
C Consulted to give input required to enable the activity to be carried out	**I** Informed of results
C^M Consulted during weekly project meeting	I^C Consult with final version prior to issue
A Accountable for ensuring the activity is correctly carried out	* An asterisk against any activity means that there is a more detailed RACI showing specific accountabilities

Names → Process/Activity ↓	Change Team Leader	Portfolio Manager			External consultant	Sponsor	Change Team	HR support
		A	B	C				
Risk Assessment 'Owner'	A	R	C^M	C^M	C^M	I	–	–
Action List 'Owner'	A	C^M	C^M	C^M	R	–	–	–
Business Process Review (gap analysis)	A	I	I	R	C	–	I	–
Business Process Design (coordination)	A	C	C	R	C	–	C	–
Business Process Design	AR*	AR*	AR*	AR*	AR*	–	C*	–
Role Descriptions Design	A	C	R	I	C	–	–	C
Change Plan Development	A	C	C	C	R	–	–	–
Change Plan Control	A	C^R	C	C	R	I	–	–
Stakeholder Management	AR	C	C	C	I	–	–	–
HR Liaison	AR	I	I	I	I	I	–	C
First Communications Development	AR	I^C	I^C	I^C	C	–	–	C
First Communications Delivery	AR	C^R	C^R	C^R	I	I	I	I
Job 'Matching'	AR	C^M	C^M	C^M	C^M	I	I	C
Simple Benefits Hierachy	A	C	C	C	R	I	I	I

Risk Table and Matrix

The risk review involved brainstorming all the potential risks which could prevent the achievement of a CSF and then analysing the probability and impact of that risk. The full Risk Table is shown in the sub-section on 'Delivery' (Stage Three page 264). However, it is simply another version of the Risk Table originally generated for the business case development (during Stage One).

During the planning stage Chris and Trish reviewed the risks and were able to check if there was additional data suggesting a risk had moved or that a different mitigation plan was needed. At this stage not many of the mitigation plans had been put into action as the project had only just been approved.

The risk summary (Table 8-11) indicated that the project looked less 'risky' than it did a month ago (during idea development) however, Trish cautioned the team not to get overly optimistic (CSFs 2 and 5 could easily go 'red').

Table 8-11 Risk summary

Critical Success Factor	**Risk rating**	
	At business case approval	**At plan approval**
CSF 1 — An Appropriate Organizational Design	Amber	Green
CSF 2 — An Engaged Change Team 'Ready for Change'	Red	Amber
CSF 3 — Robust Change Delivery Plan	Amber	Green
CSF 4 — Sponsor Support	Amber	Green
CSF 5 — Change Team Business Plan	Red	Amber
CSF 6 — Leadership	Green	Green
Overall	**Amber to Red**	**Green to Amber**

Phase 2 planning

The business processes scope area was also planned using an outline schedule, milestone plan and deliverables list using the process described in Figure 8-13.

Each individual business process (there were ten in all) was interlinked by resource, activity or decision dependencies, and these dependencies were managed at the overall project level. This allowed a greater degree of delegation to a wider team as the delivery programme progressed and therefore a greater degree of ownership of the changes.

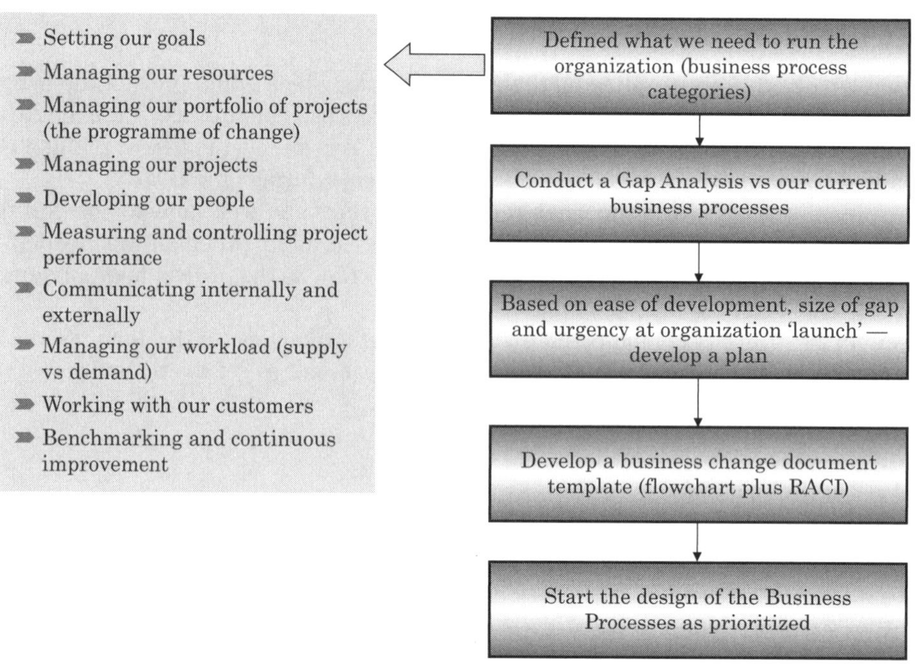

Figure 8-13 Business process design

Project delivery plan review

The plan was collated and Trish conducted a final review using the 'How?' Checklist (Table 8-12).

Table 8-12 The 'How?' Checklist

Project Management Toolkit — The 'How?' Checklist	
Project: Change Team Organizational Design	**Project Manager:** Paul Smith
Date: Month 3 Year 2	**Page:** 1 of 3
Stage One check	
Any changes since Stage One completion? (Development of the business case and project kick-off may be some time apart) No — approval was given immediately so there was no gap between Stages One and Two	
Sponsorship	
Who is the sponsor? (The person who is accountable for the delivery of the business benefits) John MacDonald **Has the sponsor developed a communication plan?** All communications activities will be managed by the Project Manager in liaison with John. He will approve the Communications Plan and then deliver the communications he needs to (mainly customers and his own management team). He will approve key messages but leave the detail to the Project Team	

(continued)

Table 8-12 (Continued)

Project Management Toolkit — The 'How?' Checklist	
Project: Change Team Organizational Design	**Project Manager:** Paul Smith
Date: Month 2 Year 2	**Page:** 2 of 3
Benefits management	
Has a Benefits Realization Plan been developed? Yes — but only in outline — adequate for now but once baseline data is developed and design completed this must be finalized **How will benefits be tracked? (Have they been adequately defined?)** A Benefits Tracking Tool has been developed so that baseline data can start to be collected now. This will help to define the exact targets for benefits metrics (timing and level)	
Business change management	
How will the business change issues be managed during the implementation of this project? (Any specific resources or organizational issues?) A change plan articulates all the issues — they are of two main types: internal Change Team and customers. BAU will be monitored as a part of the change plan to ensure that impact during the project is minimized **Have all project stakeholders been identified? (Review the stakeholder map from Stage One)** Yes — a Stakeholder Management Plan has been developed and incorporated into the Communications Plan Change readiness planning has also been completed. **What is the strategy for handover of this project to the business? (Link this to the project objectives)** Ownership will pass to the new Change Team as soon as the structure and roles are in place	
Scope definition	
Has the scope changed since Stage One completion? (Has further conceptual design been completed which may have altered the scope?) No — although there is now more detail (Table 8-9) **Have the project objectives been defined and prioritized? (What is the project delivering?)** Yes — see Table 8-9	
Project type	
What type of project is to be delivered? (For example engineering or business change) This is a Business Change Project — specifically an Organizational Design Project **What project stages/stage gates will be used? (Key milestones, for example funding approval, etc. which might be go/no go points for the project)** ➢ The plan will be approved (sponsor) and then Phase 1 design commenced ➢ Phase 1 design will need approval (sponsor) prior to implementation and prior to Phase 2 design commencing ➢ Phase 2 design is an internal approval (new organization) ➢ Project completion, handover and sustainability check will be approved by the sponsor	
Funding strategy and finance management	
Has a funding strategy been defined? (How will the project be funded and when do funds need to be requested?) A revenue budget has been developed and approved at the end of Stage One **How will finance be managed?** Usual FM Business Unit budget management processes will be followed	

(continued)

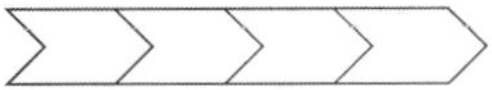

Table 8-12 (Continued)

***Project Management Toolkit* — The 'How?' Checklist**	
Project: Change Team Organizational Design	**Project Manager:** Paul Smith
Date: Month 2 Year 2	**Page:** 3 of 3
Risk and issue management	
Have the critical success factors changed since Stage One completion? (As linked to the prioritized project objectives and the critical path through the project risks) The CSF's are as originally defined but their status has changed. The project planning has reduced the overall risk profile **Have all project risks been defined and analysed? (What will stop the achievement of success?)** Yes — CSFs 2 and 5 remain the highest risks (Table 8-11) **What mitigation plans are being put into place?** Mitigation plans have been developed and can be started once this plan is approved **What contingency plans are being reviewed?** None at this time	
Project organization	
Who is the Project Manager? Paul Smith **Has a project organization for all resources been defined? (Include the Project Team and all key stakeholders)** Yes — see Figure 8-9 and Table 8-10	
Contract and supplier management	
Has a strategy for use of external suppliers been defined? (The reasons why we would need to use an external supplier for any part of the scope) Yes — require the use of one external consultant (Trish Roberts) **Is there a process for using an external supplier? (For example selection criteria, contractual arrangements, performance management)** Usual FM Business Unit procurement processes are to be followed for the duration of the contract with the external consultant	
Project controls strategy	
Is the control strategy defined? Yes ➤ Cost control strategy — revenue budget management and internal team resource management vs BAU ➤ Schedule strategy — critical milestone planning based on critical path analysis ➤ Change control — for all base design documents once approved ➤ Action/progress management — use of basic task lists vs RACI Charts ➤ Reporting — monthly one-pager to sponsor, weekly bulletins and oral report to team	
Project review strategy	
Is the Review strategy defined? (How will performance be managed and monitored—both formal and informal reviews and those within and independent to the team?) Yes ➤ **Internal — Informal** — weekly Project Team progress review meetings ➤ **Internal — Formal** — *In Control* checks every month by the external consultant ➤ **External — Formal** — Monthly review meetings with the sponsor	
Stage Two decision	
Should the Project be progressed further? (Is the project delivery strategy robust enough for project delivery to commence?) Yes — this is a robust plan which is ready for delivery	

Finally the plan was collated and Paul took the sponsor through the work before formally asking for approval to continue the project and to commence the design phase. His feedback was interesting for the team as it was an indication of how previous projects had gone straight to the delivery stage:

- He initially expected to be seeing a draft design for the organization (it was 4 weeks since he had approved the project).
- He was impressed with the level of detail and reflected on previous organizational changes which hadn't gone so well (and which hadn't been planned to this extent).
- He became even more engaged in the project as he felt this could be an organizational change case study. There were due to be many other internal reorganizations required in the coming year (linked to other business improvement projects) and having a team who not only could manage the change, but had been through it themselves, would be highly beneficial.

End of Stage Two situation

The planning stage took 4 weeks. At the end of that time the sponsor approved the plan and the communication of appropriate sections to the Change Team, the group of people who would be most impacted by the project.

Lessons learnt

A brief AAR highlighted the following key lessons learnt:

- A good project delivery plan is a positive support to future communications — ensuring clarity of key messages and addressing change resistance.
- Not all project delivery plans are the same although they all need to be able to answer the questions raised in the 'How?' Checklist.
- Keep checking that you have done enough planning (using the 'How?' Checklist) and then move on to delivery.
- Position the planning phase carefully with the sponsor. It is not unusual for very experienced business managers to challenge this phase, however the plan should easily be capable of demonstrating its benefits.

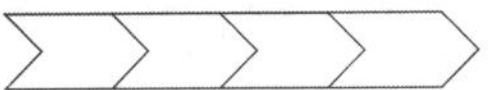

Stage Three — Project Delivery

The delivery of this change project was done in two distinct phases:

Phase 1 — design and implementation of structural changes.

Phase 2 — design and implementation of business process changes.

The team used the control tools to ensure that delivery to plan was achieved and in so doing the majority of risks were managed well. Figure 8-14 shows the outline schedule which was used and adhered to alongside more detailed tracking tools.

The team found that the Phase 1 design proceeded very smoothly with the actual design options being developed quickly due to the level of data and analysis previously conducted. However, at this stage of the project the team were mainly unaware of the changes, so the 'change impact management' was minimal.

The team quickly got into the 'project management routine' during this time:

- Tracking project activities.
- Reassessing risks.
- Reassessing culture and team 'readiness for change'.

However this was the 'calm before the storm'. Paul took the completed design to the sponsor at the end of month 3 as planned and got unanimous approval. It was then time to start internal communications.

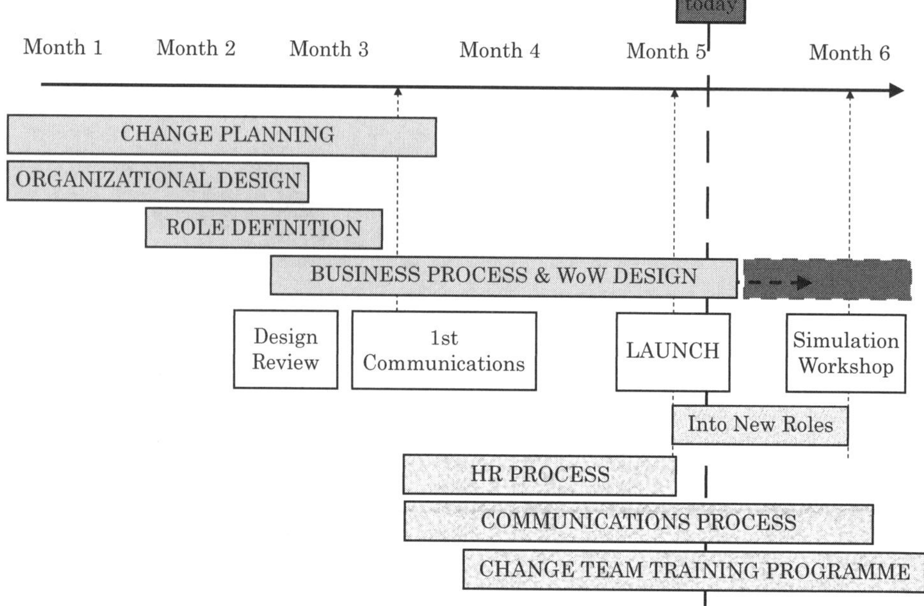

Figure 8-14 Project schedule

In readiness for this stage the team conducted a Force Field Analysis, brainstorming all the things which the Change Team might see as forces driving them to either accept or to resist the changes. As a result of this analysis the initial presentation was reviewed and the schedule developed to allow for more significant team involvement in the Business Process design, testing and implementation. At this stage the other change project in progress within the team was reviewed and incorporated into the initial communications. This again allowed a wider participation and engagement within the team.

A risk assessment (Table 8-13 and Figure 8-15) was conducted towards the end of the design phase for the initial structural changes, immediately prior to the communication of those changes to the Change Team. It forecasts that the team would be generally very receptive to the changes and agree with the logic of the design and the ultimate outcome in terms of role design. This was the general reaction at the first communication although there were one or two resistors within the team. This is very normal and Trish cautioned Paul not to over-react.

Table 8-13 Risk Table

Project Management Toolkit — Risk Table						
Project: Change Team Organizational Design				**Date:** Month 4		
Risk description			**Risk assessment**		**Action planning**	
Risk No.	**Risk description**	**Risk consequence**	**Occur?**	**Impact?**	**Mitigation plan**	**Contingency plan**
CSF1 — Appropriate Organizational Design						
1.1	Design does not meet 'must list' when put into practice	Compromises ability to deliver assignments	Low	3	Keep under review/seek feedback and support from team together with active use of the'must list' in all phases	
1.2	'Must list' we developed is flawed	Organization is wrong, compromises our ability to deliver assignments	Low	3	Keep under review/seek feed-back and support from team. Need to 'health check' the 'must list' with the team at the first communications session	
1.3	Sponsor(s) and key stakeholders do not agree with 'must list'	Lack of support, unable to deliver change	Low	4	Share benefits hierarchy with sponsor(s) and key stakeholders Use the 'must list' to explain the project outcome	
CSF 2 — An Engaged Team 'Ready for Change'						
2.1	Up to 50% of the team do not accept the need for/benefits in the OD and change programme	Team do not support the programme, become disengaged	Low	4	Assess impact on team members, set-up support network (buddy system) for those who require it	Review and decide action after first communications session

(continued)

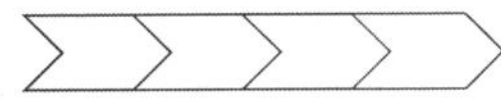

Table 8-13 (Continued)

Project Management Toolkit — Risk Table						
Project: Change Team Organizational Design				**Date:** Month 4		
Risk description			**Risk assessment**		**Action planning**	
Risk No.	**Risk description**	**Risk consequence**	**Occur?**	**Impact?**	**Mitigation plan**	**Contingency plan**
2.2	One or two team members react badly to change programme	Some team members disengage and project delivery suffers	High	4	Assess impact on team members, hold one-to-one meetings where required. Review and decide action after first communications session	Manage individual & behaviours
2.3	Unable to resource all portfolio work	Portfolio of projects not managed/tracked	Low	4	Cover this aspect of support during the design process	
2.4	Culture for change is judged as 'not right'	Unable to deliver change at all/to timescale	Low	5	Develop plans and communications strategies/ buddy systems to ensure the culture for change is right within the team	
CSF 3 — Robust Change Delivery Plan						
3.1	Change disrupts BAU	Impact on delivery leading to damage to reputation	Medium	4	Change carefully planned, communicated and rolled out. Allocate and manage resources	
3.2	Change disrupts BAU	Benefits not delivered or delivered late	Medium	4	Ensure delivery of benefits is to plan Change carefully planned, communicated and rolled out Allocate and manage resources	
3.3	Change is damaging to external perception (by customer groups)	Customers do not engage with the Change Team and perform change themselves	Low	4	Identify and articulate benefits, manage stakeholders and develop robust communication plan	
CSF 4 — Sponsor Support						
4.1	Sponsor does not support the change: he sees no benefit	Unable to proceed with change initiative	Low	5	Pre-position with sponsor, ensure the need and benefits are articulated, understood and accepted	
4.2	Leadership team do not support the change initiative	Change is difficult to position and is driven without support in place	Low	4	Position with leadership team through sponsor, ensure benefits are articulated, understood and accepted	

(continued)

Table 8-13 (Continued)

Project Management Toolkit — **Risk Table**						
Project: Change Team Organizational Design				**Date:** Month 4		
Risk description			**Risk assessment**		**Action planning**	
Risk No.	**Risk description**	**Risk consequence**	**Occur?**	**Impact?**	**Mitigation plan**	**Contingency plan**
CSF 5 — Change Team Business Plan						
5.1	People leave the group/company	Change Team under resourced and unable to deliver projects	Medium	4	Assess impact on team members, set-up support network (buddy system) for those who require it	
5.2	Greater involvement of HR or HR processes than anticipated	Need to recruit new members of the team which impacts change timescale	Medium	4	Start discussions with HR and ensure we accurately assess the impact HR is likely to have	
5.3	Team have lack of skills for new organization and new type of assignments	Unable to deliver some projects and/or consultancy assignments	High	4	Assess skills within team, analyze skills required and carry out gap analysis, produce individual development plans	Bring in external resource Match demand to team skills and capabilities
5.4	Team do not accept the need for/benefits in the OD and change programme and/or react badly to it	Team member starts an internal grievance process and additional management time is required	Low	3	Assess impact on team members, set-up support network (buddy system) for those who require it	
5.5	Lack of control of business plan	Business plan not achieved	Low	4	Regular monitoring of progress against business plan	
5.6	Lack of resource management	Inefficient use of resources	High	4	Develop resource management system/process and roll-out	
CSF 6 — Leadership						
6.1	Unexpected change in leadership	Change programme lacks direction and slows down	Low	4	Ensure all members of the leadership team fully support the programme Leaver articulates reason for leaving is not change programme but a personal opportunity	

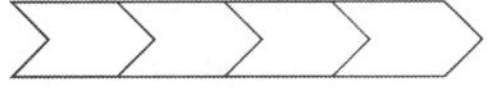

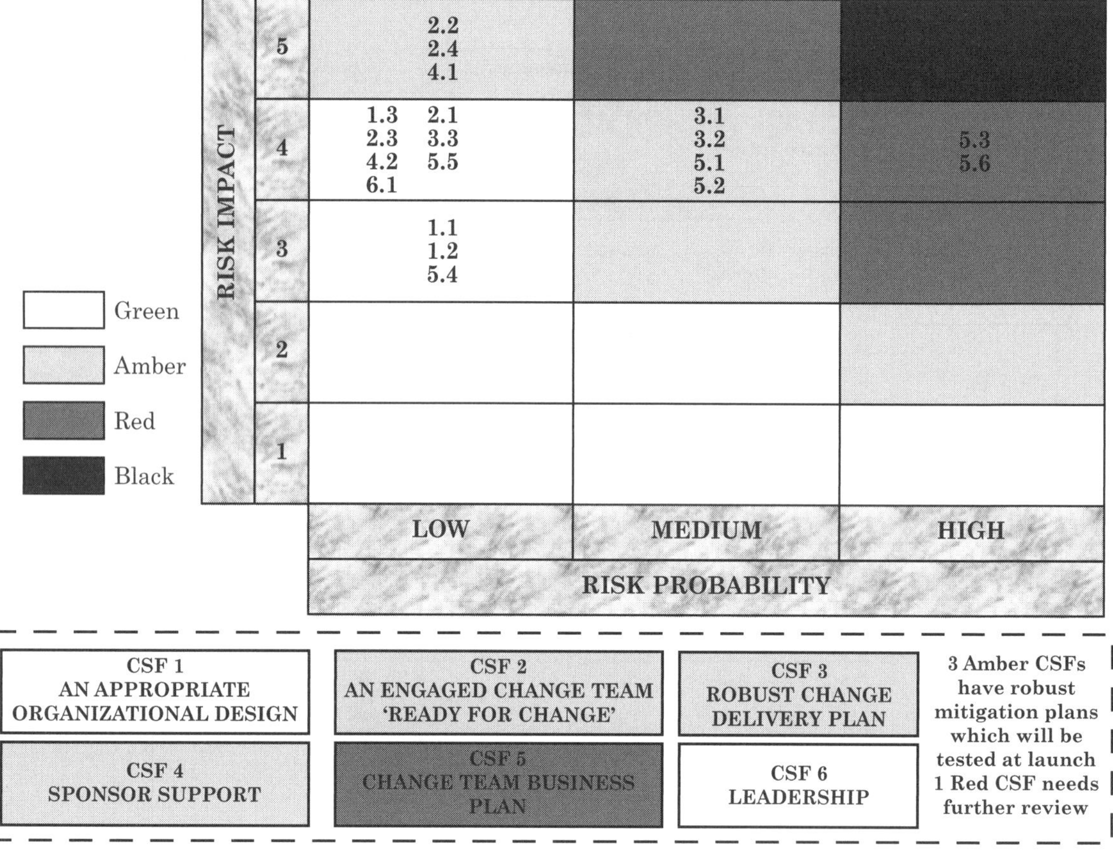

Figure 8-15 Risk matrix

The risks were analysed using the following 'scoring system'.

- Risk Probability
  - Low = <50% — unlikely to occur.
  - Medium = 50%.
  - High = >50% — likely to occur.
- Risk Impact
  - 1 = Minimal impact.
  - 2 = Minor impact on the change process.
  - 3 = Impacts one individual significantly.
  - 4 = Impacts the team significantly.
  - 5 = Stops the change programme.

After the first communication the risk profile did change slightly:

- CSF 1 remained 'green' indicating that the team were accepting the design and the risks had been appropriately mitigated.
- CSF 2 remained 'amber' with one particular area potentially turning 'red' due to the resisting behaviours of one or two team members.

- CSF 3 moved into 'amber' as it became clear after the initial communications that this change would impact Business As Usual (BAU) more than planned due to the few people issues. The mitigation plans need to be reviewed within a week or two to see if they are working.
- CSF 4 moved into 'amber' but additional positioning with the FM Business Unit management team has been scheduled.
- CSF 5 went 'red' and some contingency planning was required to ensure that the current business change projects being delivered within the FM Business Unit were not adversely impacted.
- The launch of the communications and then the implementation of the changes inevitably alters the focus of the project management. At the weekly team meetings Paul, Albert, Bill and Chris had to spend substantial time considering the current 'change climate' and specific issues within the team and within projects impacted by the changes.

However as the project progressed the Change Team became engaged within specific aspects of the Phase 2 design which gave them an element of control. There did need to be some replanning of project activities due to BAU priorities which effectively deprioritized some of the Business Processes. However, by the end of month 6 implementation was mainly complete (Table 8-14).

Table 8-14 Project Milestone Progress Report

Project Management Toolkit — Milestone Progress Report				
Project: Change Team Organizational Design		**Date:** End Month 6		
Milestone progress				
Scope area	**Activity**	**Plan (%)**	**Progress (%)**	**Comment**
1. Organizational Design — People	Formal structure — functional (designed and implemented)	100	100	Complete
	Reporting structure (designed and implemented)	100	100	Complete
	Roles and responsibilities (designed and implemented)	100	100	Complete
	HR issues resolved	100	100	Complete
	Team change issues resolved	100	100	Complete
2. Organizational Design — Process	Business process map (designed and implemented)	100	100	Complete
	Business processes designed	100	70	3 of lower priority
	Business processes implemented	100	50	Two in Month 7 One in Month 8 Two in Month 9

(continued)

Table 8-14 (Continued)

***Project Management Toolkit* — Milestone Progress Report**				
Project: Change Team Organizational Design		**Date:** End Month 6		
Milestone progress				
Scope area	**Activity**	**Plan (%)**	**Progress (%)**	**Comment**
3. Office layout	➤ Layout designed	100	100	Complete
	➤ Layout implemented	100	0	In sustainability plan
	➤ 5S (work place organization) at launch	100	100	Complete
	➤ 5S (work place organization) in normal WoW	50	0	In sustainability plan
4. Communication	➤ Project bulletins	100	100	Complete
	➤ Communications co-ordination handed to BAU	100	75	Complete in next 2 weeks
	➤ Close-out communication	100	50	
	➤ Communicate the sustainability plan	100	0	
5. Culture	➤ Culture Definition Tool designed	100	100	Complete
	➤ Culture Definition Tool implemented	75	25	In sustainability plan
6. Project performance	➤ AAR with team	0	0	AAR to be held and tracked within the sustainability plan
	➤ Project KPI review	0	50	

General Comments

Stage gate 4 (completed, approved design) achieved for Tiers 1 and 2 scope as per plan but unable to complete business process (Tier 3) design. Stage gate 5 (implementation complete) achieved for Phase 1 scope and forecast to be 1 month late for Phase 2 scope. However this is due to non-critical business process development being delayed to support BAU

Throughout the project the external consultant used the *'In Control'*? Checklist as an independent review of the project 'health'. Table 8-15 shows the health check which corresponds to the progress report in Table 8-14.

Table 8-15 'In Control?' Checklist

<table>
<tr><th colspan="2">Project Management Toolkit — The 'In Control?' Checklist</th></tr>
<tr><td>Project: Change Team Organizational Design</td><td>Project Manager: Paul Smith</td></tr>
<tr><td>Date: Month 6 Year 2</td><td>Page: 1 of 3</td></tr>
<tr><td colspan="2">Stage Two Check</td></tr>
<tr><td colspan="2">Any changes since Stage Two Completion?
Yes — the planned roll-out of the updated business processes has been changed to incorporate wider team use whilst still managing to deliver BAU</td></tr>
<tr><td colspan="2">Business change management</td></tr>
<tr><td colspan="2">What is the current status of stakeholder management? (Review the original stakeholder map and discuss)
Sponsor and FM management team — Sponsor is happy that the project has gone to plan with no HR issues
Customers of FM unit change — no impact has been reported or seen
FM business unit Project Team members — some feedback regarding the availability of Change Team members
Change Team — generally engaged with 1 or 2 individuals showing some resisting behaviours — however, it is early days in terms of 'living with the change'
How will the business be expected to operate as a result of the completion of this project?
The Change Team will operate fundamentally differently — different measures, different roles (clearer accountability and performance expectations, clearer understanding of how to work with the FM Business Unit)
Is the business ready for this project? Yes — the business needs the team to behave in their new roles so that the current projects have a step change in project performance
What is the strategy for handover of this project to the business?
The Milestone Progress Report and table of CSFs will be used to ensure that all necessary project documentation is handed over to the new organization. Some 'work in progress' will be handed over as detailed in the Month 6 progress report. This has been accepted by the new management team who have also taken ownership of the Sustainability Plan and Benefits Realization Plan. Accountability still rests with the sponsor for benefits realization but they will be tracked by the new management team</td></tr>
<tr><td colspan="2">Scope definition</td></tr>
<tr><td colspan="2">Has the scope changed since Stage Two completion?
No
What is the project progress against the Stage Two defined and prioritized project objectives?
See latest Milestone Progress Report</td></tr>
<tr><td colspan="2">Project roadmap</td></tr>
<tr><td colspan="2">What project stages/stage gates have been completed? (For example project approved, design complete)
Structure and role design has been approved (stage gate 4) and implemented (stage gate 5). Business Processes and WoW design and implementation has not achieved Stage Four entirely, although a process-by-process approval has started to allow a phased implementation</td></tr>
</table>

(continued)

Table 8-15 (Continued)

Project Management Toolkit — The 'In Control?' Checklist

Project: Change Team Organizational Design	**Project Manager:** Paul Smith
Date: Month 6 Year 2	**Page:** 2 of 3

Risk and issue management

Have all project risks been reviewed regularly during project delivery?
Yes
What is the status of mitigation plans?
At this stage the summary is: CSF 1, 3, 4 and 6 green; CSF 2 and 5 amber — all mitigation plans have delivered a reduction in the risk profile
What is the status of contingency plans?
At this stage no contingency plan has been implemented
What is the overall likelihood of achieving the project critical success factors?
There is a high probability of success and this is likely to be improved when the final issues with individual resistance and BAU workload are resolved (2 amber risks remaining out of the total risks assessed and managed)

Project organization

Are project activities being completed by the appropriate members of the organization?
Yes — team are working well together and wider Change Team has been involved in business process design

Contract and supplier management

What external suppliers are being used?
Trish Roberts is the only external supplier
What is the external supplier status and performance?
All company procedures have been followed

Project controls strategy

Are project costs under control? (Review cost plan — for example actual vs budget)
Additional resource has been used as follows:
Project Team — all team members have been used at 60% rather than 50%
External consultant — used at planned level
Change Team members — used at 25% over Phase 2 rather than planned 10%
What is the likelihood that the project budget will be maintained (forecast to completion?)
Reporting of internal resource loading is not a part of the company systems but the impact of this on other projects is seen in their benefits tracking. At this stage no benefits forecasts have been negatively impacted. The external spend will meet budget
Is the project schedule under control? (review schedule and milestone progress, etc.)
Yes — there has been some re-planning of Phase 2, but this was done in a proactive controlled manner to meet business and team needs and was entirely appropriate
What is the likelihood that the project schedule will be achieved (forecast to completion?)
The overall target was to have a sustained change by year end — this is still forecast
Are there any changes to scope (quantity, quality and functionality)? Are the costs and schedule under control?
Yes — minimal changes seen. Changes to Business Processes are all reviewed by one of the original Project Team members to check alignment with the vision and implemented structural and role changes

(continued)

Table 8-15 (Continued)

***Project Management Toolkit* — The 'In Control?' Checklist**	
Project: Change Team Organizational Design	**Project Manager:** Paul Smith
Date: Month 6 Year 2	**Page:** 3 of 3
Project review strategy	
Are regular Stage Three reviews being conducted? (Is performance being managed and monitored?) Yes — all conducted as per plan **Is project performance adequate for project success?** Yes — this has been a well-managed, agile project **Is there regular reporting? (Is the Project Team adequately managing communication of progress and performance to all stakeholders?)** Yes — there has been positive feedback from a number of stakeholders on the level and appropriateness of the reports and communications	
Stage Three decision	
Is the project under control? (Is the project control strategy robust enough for project delivery to continue?) Yes **What is the certainty that the project will be successful?** High — this project will achieve its vision	

End of Stage Three situation

The project was officially deemed to have completed implementation at the end of month 6 approximately on schedule. As the final milestone report shows, some aspects of uncompleted, non-critical scope were handed over to the new Change Team rather than delay delivery completion.

In some respects this situation always happens within an organizational design project because there is inevitably scope which is best completed by the new organization.

At the end of month 6 Paul Smith was happy that the project was effectively complete and that the new management team should take ownership.

Lessons learnt

A brief AAR highlighted the following key lessons learnt:

- It's very easy to get so involved in the control aspects of a project that you forget the people aspects. Keep using the change tools to test how people are viewing the change and how they are feeling.
- Structured project management tools are applicable to 'softer' projects which are more about people than things. If you can manage the scope of a change project more effectively then you are more capable of managing the people.

Stage Four — Benefits Realization

Once the new Change Team accepted the completed scope they started to track the benefits and the associated sustainability checks as set out in the Sustainability Plan. A sustainability plan defines three things:

- The areas of project scope which have been handed over incomplete — sustainability relies on the completion of this scope.
- Sustainability check definition — the articulation of those things which can cause or effect the realization of the benefit metrics. In effect these are the 'early warning signs' which you can see before the benefits start to decline or are non-sustained.
- Benefits tracking table — the confirmed benefit metrics with all baseline data and targets (level and date).

The management team met regularly to discuss how the new organization was progressing and they used all three areas of the sustainability plan quite overtly. In this way they were able to communicate to the wider Change Team.

This open communication of the sustainability progress, and the inevitable issues, supported the change in culture which was required to make the step change in performance.

Three months after the formal handover in month 9, the first formal sustainability check was completed. In order to ensure complete independence Paul, the Change Team Leader, once again asked Trish, the external consultant, to come in to perform the review. She completed the review through performing the following activities:

- Observing team behaviours in team meetings and in their office.
  - If you want to change culture then you need to be able to review team and individual behaviours following the change.
  - The team were used to seeing Trish around and so it was quite easy to collect this data without actually impacting the end result.
- Reviewing team performance statistics.
  - Culture is about the 'way we do things round here' and the output performance indicators are a measure of this.
  - The team had accepted the new project performance metrics and the data was very powerful in demonstrating the positive impact of the change.
- Reviewing customer feedback.
  - One of the main reasons for the project was to have a step change in project performance so that the team were better able to support their customers. Therefore any direct customer feedback is highly valuable.
  - There wasn't much of this data and more needs to be actively sought.
- Interview with the management team.
  - The new team were very open about their own performance as well as their perception of the success of the change and the performance of the team.
- One-to-one interviews with each member of the Change Team.
  - Confidential sessions between an independent reviewer and the individuals who have been impacted by the change can reveal valuable gaps in perception that can be dealt with in a confidential manner.

- The Change Team had no problem with being very open with Trish as they had done so previously on an earlier change project.
- Some still felt that the structural design could have involved them more.

- Review of the Scope Handover List.
  - Business processes — 90% were completed and 'live' by the end of month 9 (when the sustainability review took place).
- Review of the Benefits Tracking Tool.
  - Table 8-16 shows the 7 benefit metrics which are being tracked as a result of this change project
  - Figure 8-16 shows a tracking chart for one specific metric. This is a good tool for a metric which can be measured regularly over time
  - Figure 8-17 is a Radar Chart which has taken all benefits and analysed each against the plan for month 9 so that the benefit gaps can be seen as well as areas of high performance

Table 8-16 Benefits Tracking Tool

Project Management Toolkit — Benefits Tracking Tool

Project: Change Team Organizational Design | **Date:** Month 9

Benefit metric		Baseline As at start of the project	Milestone 1 Once change implemented (End M6)	Milestone 2 2nd Month (End M7)	Milestone 3 3rd Month (End M8)	Target 6th months implementation (End M12)
Change Team utilization (% of time on core projects)	Plan Actual	60% 60%	60% 54%	65% 62%	70% 69%	80%
Percentage of projects progress to plan	Plan Actual	50% 50%	70% 80%	75% 75%	80% 80%	90%
Portfolio speed (projects/year)	Plan Actual	5 5	Unable to measure within this timescale			10 10
Cycle time — business case approval (0–2)	Plan Actual	16 weeks 16 weeks	Only tracked each quarter		12 weeks 10 weeks	8 weeks
Cycle time — approved plan (Phase 3)	Plan Actual	2 weeks	Only tracked each quarter		4 weeks 4 weeks	4 weeks
Cycle time — delivery (4 and 5)	Plan Actual	35 weeks 35 weeks	Only tracked each quarter		28 weeks	22 weeks
Percentage of benefits delivered to plan	Plan Actual	50% 50%	60% 60%	65% 80%	75% 90%	90%

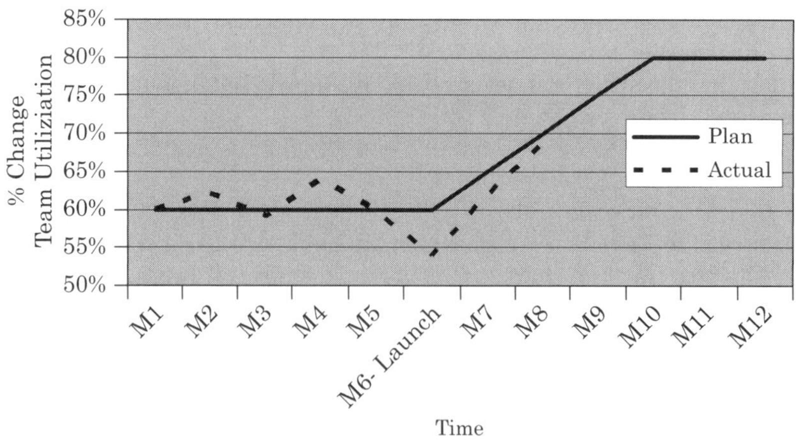

Figure 8-16 Benefits tracking chart — Change Team Utilization

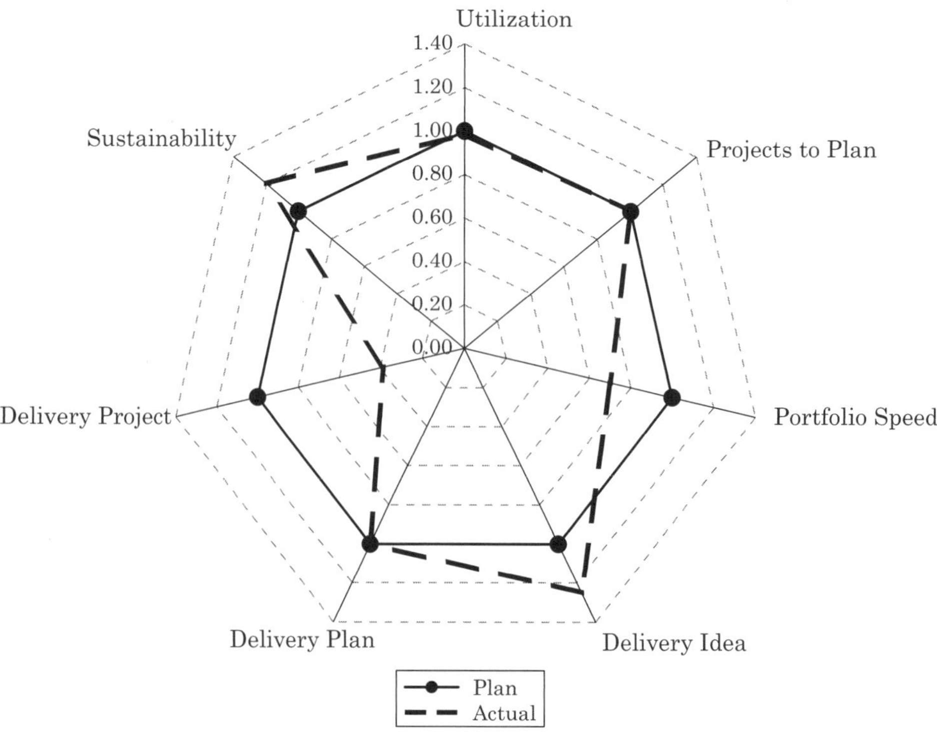

Figure 8-17 Benefits radar chart — summary benefits assessment

Based on the above reviews the results are collated and summarized in the following documents:

- The Sustainability Checklist (Table 8-17)
- The Benefits Realized Checklist (Table 8-18)

The key sustainability checks tried to identify the things which would be precursors of the benefits:

- Improved project performance required a change in behaviours — more positive, more motivated and more informed. The sustainability checks, looking at the way the team behaved in the office, were a good sign that they had changed their behaviours.
- Improved Change Team performance (seen in the reduction in variation of project outcomes) required more formal business processes in place.

Table 8-17 Sustainability Checklist

Project Management Toolkit — Sustainability Checklist				
Project: Change Team Organizational Design			**Date:** Month 9	
Project vision				
Once this project has been successfully completed, and all changes sustained, the Change Team will be better positioned to deliver its critical path of success and objectives for the coming year and beyond. It will reinforce the sustainability of the changes in Year 1 and introduce new WoW which will sustain the Change Team, and the benefit it can deliver, for the FM Business Unit. This vision of sustainability will be evident through the team's actions, behaviours and general culture and attitudes. Use of new business processes and maintenance of the new office layout, in terms of how it can positively support the sustainability of the change, should be evident. The group management should positively support the sustainability.				
Sustainability review information				
Previous sustainability review: End of project review			**This sustainability review:** Q1 after implementation	
Project representative: Project Manager			**Customer representative:** Change Team	
Sustainability checks				
Check No.	**Check**	**Target** (sustained change)	**Last review**	**This review**
1.	Office layout	In place and no changes	No	Office relocated and layout done per plan and maintained
2.	Office culture	Office tidy and adhere to ground rules	Existing office cleared out	Team have developed ground rules to support more effective office working and these appear to be "in use" — office is tidy
3.	Office Communications	Extensive use of visual communications	None	Team vision and CSFs, ground rules and portfolio tracking system all on wall A professional (rather than social) culture is the expected norm

(continued)

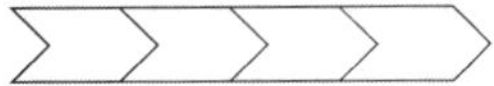

Table 8-17 (Continued)

Project Management Toolkit — Sustainability Checklist

Project: Change Team Organizational Design	**Date:** Month 9

Sustainability checks

Check No.	Check	Target (sustained change)	Last review	This review
4.	Business Processes 'live'	100% in place and 'live' by month 6	70% developed and 50% 'live'	90% in place and 'live' — one still to complete
5.	Business Processes used	90% compliance by month 12	25% compliance of 50% 'live'	75% compliance is good at this stage
6.	Team time diaries	Team actively measure utilization	No	Utilization is still completed through an ad hoc process and individuals are still not engaged in this. Strong management lead is required Team are managing their time better but still are not able to estimate accurately
7.	Team motivation	High	Neutral	Medium and growing. The one-to-ones showed that the team are starting to see the benefits of this change for themselves
8.	Change Team meet Year 2 objectives	100% by month 12	50% red; 50% amber	25% green; 70% amber; 5% red
9.	Process to bring new ideas into the portfolio	Flow appropriate	None	Steering process is in place but portfolio is full and no new ideas required
10.	Customer feedback	1 unsolicited positive feedback per month	None	None — however it is early days in terms of impact on the current project outcomes
11.	Decreased external 'helpdesk' support	To zero by end Year 2 (related to supporting the team)	1 day/week to close out the project	1 day/month from month 9 onwards to zero at month 12

Summary comments and next steps

At this stage the changes appear to be progressing to a sustainable state and those from this project are being supported by the changes delivered from the Change Process Improvement Project.

Is the change completely sustained?	~~Yes~~/No	**Date of next sustainability check**	End of next quarter

Table 8-18 'Benefits Realized?' Checklist

<table>
<tr><th colspan="2">Project Management Toolkit — The 'Benefits Realized?' Checklist</th></tr>
<tr><td>Project: Change Team Organizational Design</td><td>Project Manager: Paul Smith</td></tr>
<tr><td>Date: Month 9 Year 2</td><td>Page: 1 of 1</td></tr>
<tr><th colspan="2">Stage Three check</th></tr>
<tr><td colspan="2">Any changes since Stage Three completion? (Note only the changes since the final Stage Three 'health check')
Yes — team has been physically relocated to a new building and layout design was altered but used original concepts</td></tr>
<tr><th colspan="2">Business benefits</th></tr>
<tr><td colspan="2">Has the business case changed since Stage One? (For example during planning and delivery, pre- or post-project approval)
No — the Benefits Hierarchy which was approved at the start of the project remains valid along with the Benefits Specification Table and associated Business Case
Have all benefits been defined in terms of trackable metrics? (Why is the project being done?)
Yes — the key benefits of this project have been converted to trackable metrics. Some will only be measurable within a longer time frame as they are linked to project lead-time. Currently the lead-time is longer than the ideal tracking milestone frequency. This should change as the cycle time reduces and project performance increases
What is the customer feedback? (Feedback from all stakeholders in the business including the customer)
Sponsor — appears to be a successful project — keen to see evidence in the benefits tracking at month 12
FM business unit managers — hasn't impacted them or their projects although one team member did have to wait for one idea to be resourced which caused some issues at the time
Change Team members — generally highly engaged in the new WoW although culture change does take time and individual behaviours need to be tracked and assessed on a frequent basis
Are the benefits being tracked?
Yes — see Table 8-16, Figures 8-16 and 8-17</td></tr>
<tr><th colspan="2">Business change</th></tr>
<tr><td colspan="2">Is the business ready for this project? (If the project can only enable benefits delivery by changing the way people work — has this been delivered, for example training?)
The team did need the associated process training (an output of the parallel change project: Change Process Improvement) which reinforced some of the culture changes within the structural and role changes. Additionally the eventual layout changes were better than anticipated as there was a complete relocation. This allowed a completely new WoW which positively reinforced the new culture</td></tr>
<tr><th colspan="2">Scope definition</th></tr>
<tr><td colspan="2">Has the scope been delivered?
Yes — although some handover scope is still being completed (non-critical)
Have the benefit enablers been delivered? (Are you sure that the project will enable the benefits to be delivered now the project is complete?)
The sustainability checks confirm that in the main all enablers are in place. The sustainability checking will continue until it is clear that the benefits are sustained</td></tr>
<tr><th colspan="2">Stage Four decision</th></tr>
<tr><td colspan="2">Has the project been delivered? (Delivery of project critical success criteria)
Yes — although (BAU) work was impacted at launch and the team did stop accepting new projects into the portfolio for 2 months
Have the business benefits been delivered? (Why was the project done in the first place?)
No — the benefits won't be completely delivered until month 12; however, signs are positive that this will be the case</td></tr>
</table>

End of Stage Four situation

Following the sustainability review the Project Team was formally 'closed' and the new management team (led by Paul Smith) continued to track benefits and to also discuss and review areas of sustainability.

- The team were more focused on the benchmark delivery times and used these actively to estimate project cycle times and milestones. As a result plans were more robust and were adhered to.
- Individual Change Managers had fewer projects at any one time, but they went through the portfolio faster.
- Project delivery time was the most variable benefit metric, but from a root cause analysis the team are not concerned due to the great variety of business change that they are involved in.

Paul felt that he was now more able to make the 'right' decisions and in doing so support the FM Business Unit better.

Figure 8-18 shows the IPO for the close-out session with the whole Change Team (held prior to the AAR). This celebration of the completion of the change was critical in moving the team out of 'project mode' and back into BAU

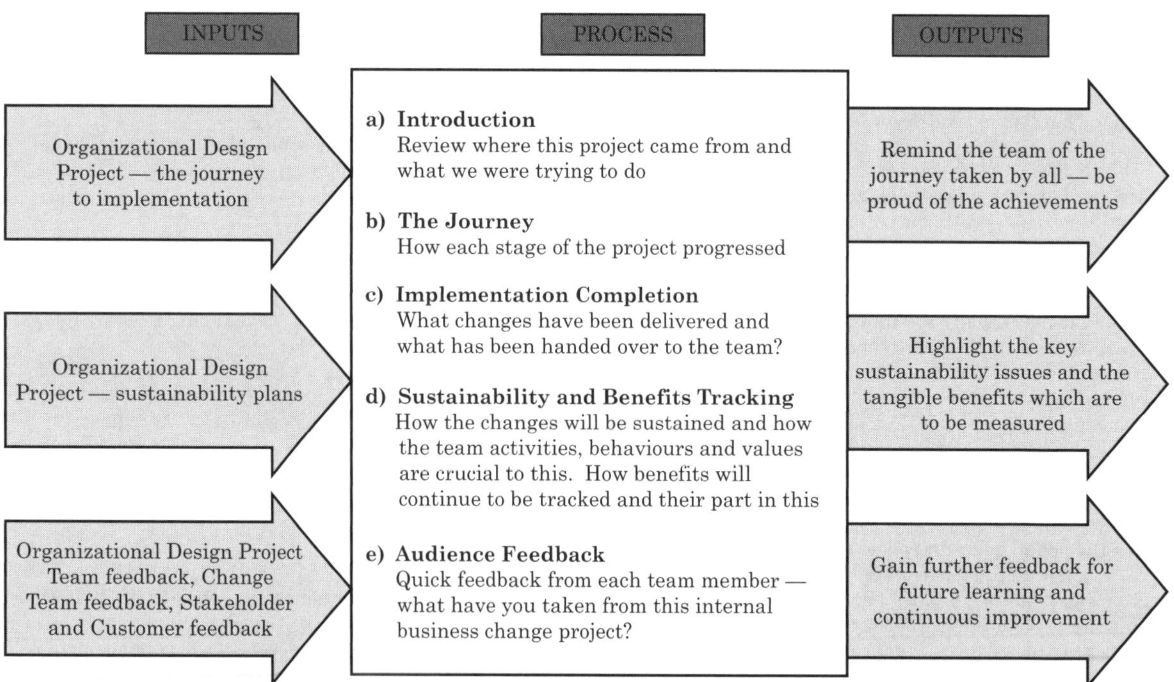
INPUTS
PROCESS
OUTPUTS
Organizational Design Project — the journey to implementation
Organizational Design Project — sustainability plans
Organizational Design Project Team feedback, Change Team feedback, Stakeholder and Customer feedback
a) Introduction
Review where this project came from and what we were trying to do
b) The Journey
How each stage of the project progressed
c) Implementation Completion
What changes have been delivered and what has been handed over to the team?
d) Sustainability and Benefits Tracking
How the changes will be sustained and how the team activities, behaviours and values are crucial to this. How benefits will continue to be tracked and their part in this
e) Audience Feedback
Quick feedback from each team member — what have you taken from this internal business change project?
Remind the team of the journey taken by all — be proud of the achievements
Highlight the key sustainability issues and the tangible benefits which are to be measured
Gain further feedback for future learning and continuous improvement

Figure 8-18 IPO diagram for the Change Team close-out session

During the formal AAR, attended by all members of the Change Team, the "end of AAR quotes" reinforced the view that this was a successful and well-run project. It was not without its 'ups and downs' but was always in control.

The fact that the project was able to achieve the external benchmark delivery cycle times for all phases in the Business Change Project Roadmap (Figure 8-3) was a great way to 'model' the new behaviours expected from the Change Team. Additionally the team were able to gain knowledge and additional learning points from the project which were integrated into their own change project delivery process.

Lessons learnt

A brief AAR highlighted the following key lessons learnt:

- A sustainability plan, like any other form of plan, is only useful if you use it — therefore checking progress is a value-added activity for any management team.
- Benefits tracking is a valuable operational activity, and if you have identified the right benefits then it is usual for the benefits to become standard operational measures.
- Sustainability checks should not become operational measures as a matter of norm, so make sure that you understand the difference between sustainability checks and benefit metrics.
- Closing out a project is an important phase for all project participants — as well as allowing reflections on success it also moves everyone back into 'business mode'.
- A formal AAR (as distinct from a close-out session) is needed if all the good and poor parts of the project are to be used positively within the organization. Build the good things into the 'normal' project process and ensure that the poor things are not able to occur again.

Conclusions

Every project has the potential to change the business into which it is being implemented. Often the challenge for a Project Manager is to identify the potential changes, either required or as a direct result of the project.

In a 'pure' business change project the challenge is to recognize and articulate the tangible things that the project will deliver and the associated business benefits. In this way business change 'adds value' to the organization.

Many projects, in a variety of industries, are focused on business change as a way to operate better and to deliver customer and/or shareholder value. These projects can be:

- Process improvements (services, manufacturing or supply chain) — where waste is removed, efficiency improved and variation reduced.
- Organizational changes — driven by acquisitions and mergers as well as by organizational issues.
- Organizational relocations — moving operations or people to rationalize an organization, or as a part of a merger/acquisition/redesign.

The aim of including such a project in this book is to demonstrate the application of *Project Management Toolkit* outside of the traditional engineering or technology project.

The team involved with this case study actively used *Project Management Toolkit* alongside their change toolkit and Business Change Project Roadmap. They gave substantial feedback on each of the tools and a sample of this feedback is contained in Table 8-19.

The team understood how to effectively use both project and change tools to support the change cycle (Figure 8-19) which:

- Would occur within their Change Team on a continuous basis as they improved the way they supported the business
- Would occur within every project they delivered for the business.

Table 8-19 Sample Project Management Toolkit Feedback

Tool	Stage used	Feedback
➢ Simple Benefits Hierarchy	Mainly Stage One but initially on projects in delivery to prove we shouldn't be doing them	➢ This tool fundamentally changed the way that projects entered our change portfolio ➢ We had a way of saying NO to projects which previously wouldn't have 'felt right'; this tool gives us a clear alignment test that cannot be argued against
➢ The 'Why?' Checklist	Mainly Stage One but initially on projects in delivery to prove we shouldn't be doing them	➢ Even with an aligned project it was still possible to rush to the delivery stage and this tool gave Project Managers 'pause for thought' ➢ If we couldn't answer these basic questions we knew that we didn't know enough about our project
➢ The 'How?' Checklist	Mainly Stage Two but initially on projects in delivery to prove we shouldn't be doing them	➢ Before I used this tool I thought that I was a good planner — but then I realized that my schedule did not explain how I would deliver each aspect of the project ➢ I tested this on a number of my previous projects and now understand why they didn't deliver as expected — they didn't have a robust plan, in fact we went straight from approval to design ➢ A great tool to sanity check a project before delivery
➢ The 'In Control?' Checklist	Stage Three	➢ You can't hide from the questions in this checklist — it really makes you challenge whether you do know what's going on in your project
➢ The 'Benefits Realized?' Checklist	Stage Four	➢ A great tool to support handover of a completed project to the business user
➢ Benefits maps, specification and tracking tables	Used in all stages	➢ Starting to think about benefits metrics in the early stages meant that we had time to collect the baseline data and then track the changes as the project was implemented ➢ It made us really think hard about what could and couldn't be measured and helped us to understand our business
➢ Critical Path of Success and associated Risk Table and Matrix	Used in all stages	➢ Until we had this tool we all delivered a project like a standard 'recipe' without considering the specific needs of a project to achieve the vision of success ➢ This tool made us articulate the critical aspects, without which we wouldn't achieve success ➢ The use of the Risk Table also made us take a step back and evaluate project progress a completely different way. We still used traditional progress tools and then used this to assess if we 'stood a chance' of achieving success ➢ Our focus on risk management has changed
➢ Sustainability Checklist	Used in all stages	➢ Fundamentally understanding how to sustain change in a business context is a priceless skill ➢ This tool helped us to understand that initially we didn't have this skill: we measured project objectives and we tracked benefit metrics but we never looked at the things which supported the sustainability of the latter or how we could build it into the former ➢ It helped us to build a real vision of change at a very early stage in a project — this was a great tool for stakeholder management

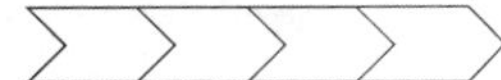

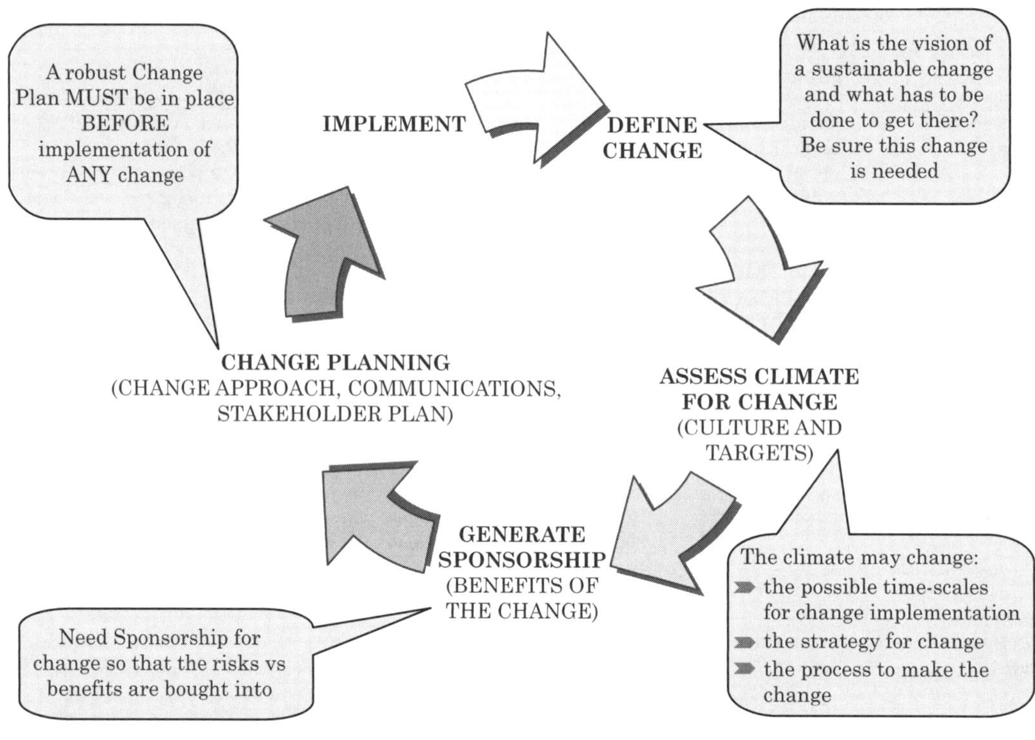

Figure 8-19 The Change Cycle

And finally...

- The tools in this book are as applicable to business change projects as they are to other project types such as engineering or IT.
- *Project Management Toolkit* is not exhaustive nor is it meant to be used in a prescriptive manner:
  - Select the tools to use based on the needs of the project, the business and the Project Manager.
  - Supplement your toolkit with other tools as necessary to support project delivery success.
- The four stage checklists can be used as a reminder to 'step back' from the detail of a project and objectively answer the questions:
  - *Why* are we doing this project?
  - *How* should we deliver this project?
  - Are we *In Control* of this project?
  - Have *Benefits* been *Realized?*
- Use of the toolkit requires data and people:
  - High quality, objective data
  - Appropriate interactions with the appropriate stakeholders, customers, sponsor and/or Project Team members
  - The tools can only support delivery of a quality analysis or outcome if they have appropriate inputs: garbage *In* = garbage *Out*

Index